CONTENTS

APPENDIX J—INVESTIGATIONS

APPENDIX K—NONLETHAL WEAPON PROCEDURES

APPENDIX L—ARREST/APPREHENSION/DETAINMENT

LIST OF ILLUSTRATIONS

PREFACE

NTTP 3-07.2.3, AUG 2011 LAW ENFORCEMENT AND PHYSICAL SECURITY, is a reference for regional commanders, regional security officers, commanding officers, security officers, administrative staffs, and Navy security forces.

This publication provides tactics, techniques, and procedures governing the conduct of physical security and law enforcement. It outlines Navy policies and objectives and includes tools to help organize, plan, train for, and implement effective and efficient physical security and law enforcement programs. Although focused on Navy installations, the techniques and procedures within this document have applicability to afloat and expeditionary operations. Unless otherwise stated, masculine nouns and pronouns do not refer exclusively to men.

Report administrative discrepancies by letter, message, or e-mail to:

COMMANDER
NAVY WARFARE DEVELOPMENT COMMAND
ATTN: DOCTRINE
1528 PIERSEY STREET BLDG O-27
NORFOLK VA 23511-2723

NWDC_NRFK_FLEETPUBS@NAVY.MIL

CHANGE RECOMMENDATIONS

Procedures for recommending changes are provided below.

WEB-BASED CHANGE RECOMMENDATIONS

Recommended changes to this publication may be submitted to the Navy Doctrine Library System, accessible through the Navy Warfare Development Command website at: http://ndls.nwdc.navy.smil.mil or https://ndls.nwdc.navy.mil.

URGENT CHANGE RECOMMENDATIONS

When items for changes are considered urgent, send this information by message to the Primary Review Authority, info NWDC. Clearly identify and justify both the proposed change and its urgency. Information addressees should comment as appropriate. See accompanying sample for urgent change recommendation format on page 27.

ROUTINE CHANGE RECOMMENDATIONS

Submit routine recommended changes to this publication at any time by using the accompanying routine change recommendation letter format on page 28 and mailing it to the address below, or posting the recommendation on the Navy Doctrine Library System site.

COMMANDER
NAVY WARFARE DEVELOPMENT COMMAND
ATTN: DOCTRINE
1528 PIERSEY STREET BLDG O-27
NORFOLK VA 23511-2723

CHANGE BARS

Revised text is indicated by a black vertical line in the outside margin of the page, like the one printed next to this paragraph. The change bar indicates added or restated information. A change bar in the margin adjacent to the chapter number and title indicates a new or completely revised chapter.

WARNINGS, CAUTIONS, AND NOTES

The following definitions apply to warnings, cautions, and notes used in this manual:

WARNING

An operating procedure, practice, or condition that may result in injury or death if not carefully observed or followed.

CAUTION

An operating procedure, practice, or condition that may result in damage to equipment if not carefully observed or followed.

Note

An operating procedure, practice, or condition that requires emphasis.

WORDING

Word usage and intended meaning throughout this publication are as follows:

"Shall" indicates the application of a procedure is mandatory.

"Should" indicates the application of a procedure is recommended.

"May" and "need not" indicate the application of a procedure is optional.

"Will" indicates future time. It never indicates any degree of requirement for application of a procedure.

FM ORIGINATOR

TO *(Primary Review Authority)*//JJJ//

INFO COMNAVWARDEVCOM NORFOLK VA//

COMUSFLTFORCOM NORFOLK VA//JJJ//

COMUSPACFLT PEARL HARBOR HI//JJJ//

(Additional Commands as Appropriate)//JJJ//

BT

CLASSIFICATION//N03510//

MSGID/GENADMIN/*(Organization ID)*//

SUBJ/URGENT CHANGE RECOMMENDATION FOR *(Publication Short Title)*//

REF/A/DOC/NTTP 1-01//

POC/*(Command Representative)*//

RMKS/ 1. IAW REF A URGENT CHANGE IS RECOMMENDED FOR *(Publication Short Title)*

2. PAGE _____ ART/PARA NO _____ LINE NO _____ FIG NO _____

3. PROPOSED NEW TEXT *(Include classification)*

4. JUSTIFICATION.

BT

Message provided for subject matter; ensure that actual message conforms to MTF requirements.

Urgent Change Recommendation Message Format

DEPARTMENT OF THE NAVY
NAME OF ACTIVITY
STREET ADDRESS
CITY, STATE XXXXX-XXXX

5219
Code/Serial
Date

FROM: *(Name, Grade or Title, Activity, Location)*
TO: *(Primary Review Authority)*

SUBJECT: ROUTINE CHANGE RECOMMENDATION TO *(Publication Short Title, Revision/Edition, Change Number, Publication Long Title)*

ENCL: *(List Attached Tables, Figures, etc.)*

1. The following changes are recommended for NTTP X-XX, Rev. X, Change X:

   a.   CHANGE: (Page 1-1, Paragraph 1.1.1, Line 1)
Replace "...the ~~National Command Authority~~ President and Secretary of Defense establishes procedures for the..."
REASON: SECNAVINST ####, dated ####, instructing the term "National Command Authority" be replaced with "President and Secretary of Defense."

   b.   ADD: (Page 2-1, Paragraph 2.2, Line 4)
Add sentence at end of paragraph "See Figure 2-1."
REASON: Sentence will refer reader to enclosed illustration.
Add Figure 2-1 (see enclosure) where appropriate.
REASON: Enclosed figure helps clarify text in Paragraph 2.2.

   c.   DELETE: (Page 4-2, Paragraph 4.2.2, Line 3)
Remove "Navy Tactical Support Activity."
"...~~Navy Tactical Support Activity, and~~ the Navy Warfare Development Command ~~are~~ is responsible for..."
REASON: Activity has been deactivated.

2. Point of contact for this action is *(Name, Grade or Title, Telephone, E-mail Address)*.

(SIGNATURE)
NAME

Copy to:
COMUSFLTFORCOM
COMUSPACFLT
COMNAVWARDEVCOM

Routine Change Recommendation Letter Format

CHAPTER 1

Introduction

1.1 PURPOSE

This publication provides tactics, techniques, and procedures (TTP) governing the conduct of physical security (PS) and law enforcement (LE) at Navy installations, within expeditionary forces and onboard Navy ships. It provides a basis for understanding Navy policies and objectives related to PS and LE. More important, it provides regional commanders (REGCOMs), regional security officers (RSOs), installation and ship commanding officers (COs), security officers (SO), administrative staffs, and the Navy security force (NSF) with needed tools to help organize, plan, train for, and implement effective and efficient PS and LE programs using the limited resources at their disposal. Both PS and LE programs include measures taken by a command, ship, or installation to protect against all acts designed to, or that may, impair its effectiveness. In other words, both PS and LE personnel provide security and are key to the protection construct. Within this document, installation security officers and afloat security officers will be referred to as SOs.

The guidance within this publication will assist the REGCOMs, RSOs, COs, and SOs to efficiently allocate security resources, effectively employ the NSF, and expeditiously increase the response to incidents within a Navy installation. Key to ensuring an effective and efficient PS/LE program is the integration of all security and law enforcement forces at an installation into a single force structure.

Antiterrorism (AT) as it applies to installations is also included within the protection construct. This Navy Tactics, Techniques, and Procedures (NTTP) is concerned with only PS and LE. For TTP that support the Navy AT mission, refer to NTTP 3-07.2.1, Antiterrorism.

The primary missions within the PS construct are to establish a force protection perimeter, including assessment zone, warning zone, threat zone, and the minimum standoff zone. This task includes enforcing perimeter boundaries and conducting access control to prevent unauthorized entry and counter the introduction of unauthorized personnel, hazardous materials (HAZMAT), contraband, and prohibited items from entering an installation or restricted area. Paramount within this task is the requirement to provide NSF for the protection of vessels and port/waterfront facilities, including protecting friendly forces within a designated geographic area, harbors, approaches, or anchorages against external threats, sabotage, subversive acts, accidents, theft, negligence, civil disturbance, and disasters.

The primary missions for LE on a Navy installation are to enforce laws and regulations that maintain the discipline of units and personnel, preserve the peace, protect the life and civil rights of persons, and provide first response. LE includes a visible deterrent to threats; the initial response to security-related incidents; investigation of incidents; a valuable resource for the collection, processing, and dissemination of criminal intelligence; and formal liaison with civilian and other military LE agencies. Included in this mission is understanding the basis for the standing rules of engagement (ROE) and standing rules for the use of force (RUF) and how to apply the use-of-force continuum, including use of deadly force.

In some instances, the distinction between police and guard work may not be an easy one to make because of the similarities between the two kinds of work and because, in many cases, the two functions are performed by the same individual at the same time. Both police and guards are armed, they usually are uniformed, and all are trained to respond to similar situations, referred to or addressed as officer, and subject to substantial hazard or danger in emergencies. LE personnel are typically trained to respond to misdemeanors and felonies, which can range from petty theft and verbal assault through murder and other conditions involving violations of law and threats to human life. Security guards are trained more in the methods and techniques for detecting and repelling attempts at trespass, sabotage, and theft of property. Typically, security guards maintain situational awareness; sound the alarm, react to, and/or resist attempted violations; detain offenders; and turn over cases and violators to police or other LE officers. Security guards generally have strict limitations in responding outside their assigned post with their authority and responsibilities clearly defined in post orders. Many of the skills required for both PS and LE are in the techniques and procedures outlined in this NTTP.

This NTTP is not all-inclusive, nor does it relieve the CO and SO from their obligation to comply with statutory and regulatory requirements.

1.2 SCOPE

This publication is applicable to all U.S. Navy installations, ships, and units within the United States or foreign territory in accordance with (IAW) applicable status-of-forces agreements (SOFAs). This NTTP provides required procedures for the protection of Navy installations and supports the requirements established in OPNAVINST 5530.14 series, Navy Physical Security and Law Enforcement Program. Although focused on Navy installations, many of the techniques and procedures within this document also apply to afloat and expeditionary operations.

1.3 PHYSICAL SECURITY OVERVIEW

PS is that part of security concerned with physical measures designed to safeguard personnel; to prevent unauthorized access to equipment, installations, materiel, and documents; and to safeguard them against espionage, sabotage, damage, and theft. PS involves the total spectrum of procedures, facilities, equipment, and personnel employed to provide a secure environment. The essence of PS on Navy installations, at locations where military personnel reside, and during in-transit operations involves the integration of policy, doctrine, personnel, materiel, training, intelligence, and planning. This integration achieves the maximum return on investment while allowing the CO to accept certain risks yet capitalize on optimum countermeasures.

Viewing installation security as a three-layered environment or layered defense focused on major security missions improves protection from purely physical to full-spectrum. The operational construct of the battlespace for an CO is based on the concept of the operation, accomplishment of the mission, and protection of the force. The CO uses experience, professional knowledge, and understanding of the situation to change his battlespace as current operations transition to future operations. The installation boundary is the area of operations (AO), consisting of an inner ring that must be protected, such as facilities, ships, aircraft, or activities within an installation that, by virtue of their function, are evaluated by the CO as vital to the successful accomplishment of the installation mission. Also part of the AO or installation is the middle ring, which includes the rest of the installation confines, such as housing, ammunition storage areas, non–single-point-of-failure communications sites, etc. The outer or third ring does not have a definite boundary and is defined by the CO as the area of influence. The installation area of influence includes areas that the CO can influence through such activities as contacts with local officials, LE agencies, and emergency management (EM) agencies and through command public affairs officers (PAOs).

This NTTP specifically addresses those techniques and procedures that support full-spectrum security, such as PS assessments and loss prevention. Many of the other functions within PS are covered in NTTP 3-07.2.1, Antiterrorism, as many aspects of AT and PS overlap.

1.4 LAW ENFORCEMENT OVERVIEW

Within Navy installations, NSF conduct extensive law-and-order operations. These operations support order and discipline on installations and ships around the world. The NSF professional orientation ensures their ability to operate with restraint and authority and with a minimum use of force, making the NSF ideally suited as a response force in every incident of disorderly conduct. Wherever Navy installations or ships are located, NSF are the protectors and assisters of the military community. NSF efforts ensure a lawful and orderly environment. By their efforts to preserve the law, NSF help the commander ensure a high standard of order and discipline within the commander's units. This NTTP specifically addresses those techniques and procedures that support the conduct of LE operations on Navy installations and onboard Navy ships.

CHAPTER 2

Organization and Responsibilities/Operating Environment

2.1 OVERVIEW

PS and LE are primary missions of the NSF, coupled with AT for the activity to which assigned. The primary mission and purpose of PS and LE personnel are to protect the installation, ship and property, enforce law, maintain law and order, preserve the peace, and protect the life and civil rights of persons. The NSF consists of armed personnel regularly engaged in LE and security duties involving the use of deadly force, and management and support personnel who are organized, trained, and equipped to protect Navy personnel and resources under Navy authority. The NSF consists of the Master-at-Arms (MA), active and reserve personnel, the auxiliary security force (ASF), and Department of the Navy (DON) civilian personnel and contract guards. Contract guards and ASF may not perform LE duties; the function of ASF and contract guards is physical security. The PS and LE element of the NSF shall endeavor to:

1. Protect life, property, and information.

2. Conduct LE patrols, including military working dogs.

3. Respond to alarms and calls for service.

4. Enforce laws, rules, regulations, and statutes.

5. Secure the scene at incidents and provide first aid and take other immediate actions at scenes to preserve life and to prevent further injury.

6. Control traffic.

7. Effect detentions and apprehensions as needed.

8. To the extent necessary, process crime scenes and gather evidence.

9. Conduct criminal and accident investigations.

10. Prepare and submit incident reports (IRs) and other related correspondence.

11. Testify in judicial and administrative proceedings.

12. Issue Administrative Citations (DD 1408s) and/or U.S. Magistrate Central Violation Bureau form.

13. Conduct specialized LE programs required by location/mission of the command, such as wildlife protection.

14. Provide escorts, which may include vessels, funds, personnel, etc.

15. Respond to noncriminal service requests (roadside assistance, etc.).

16. Educate command personnel in crime and loss prevention programs.

17. Maintain awareness of all local threats.

18. Liaison with other military and civilian law enforcement agencies.

19. Assist civilian agencies when called on to do so IAW the Posse Comitatus Act (18 U.S.C. 1385).

2.2 ORGANIZATION

The NSF on installations is organized in a construct, where consolidated divisions (i.e., military working dog (MWD), administration, harbor security, training, and investigations) have been established, with regional staff oversight and responsibility for training; resourcing; and force protection, PS, and LE policy oversight. (See figure 2-1 for the organization of the NSF.) Tactical control rests with the CO, who has the responsibility for the safety and security of the command through the SO.

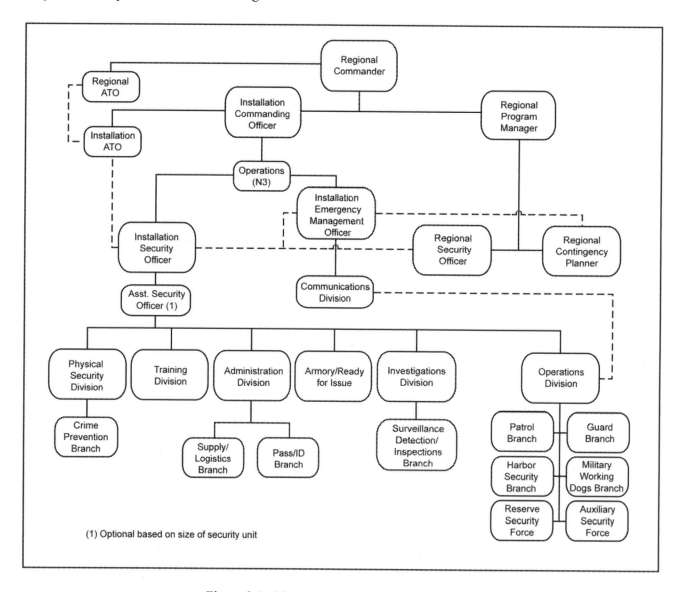

Figure 2-1. Navy Security Force Organization

2.3 COMMANDING OFFICER

The CO is responsible for protecting installation and ship personnel and assets, as well as maintaining good order and discipline. To accomplish this mission, the CO provides tactical direction to the SO.

2.4 SECURITY OFFICER

The SO is the principal staff officer to the command for PS and LE matters, with responsibility for the following functional areas.

1. Supervising and providing leadership within each of the division and branch functional areas within the security department.

2. Managing the performance of the following administrative duties and shall be prepared to perform any of the associated tasks listed below:

    a. Maintain records and correspondence.

    b. Retain and manage office security records as provided by reference SECNAV M-5210.1, Department of the Navy (DON) Records Management Manual, and SECNAV M-5210.2, Department of the Navy (DON) Standard Subject Identification Code (SSIC) Manual.

    c. Manage the Department of the Navy Consolidated Law Enforcement Operations Center (CLEOC) installation LE element.

    d. Manage PS waiver and exception program.

        (1) Identify the requirement for a PS waiver or exception.

        (2) Research historical waivers or exceptions.

        (3) Identify mitigations/compensatory measures.

        (4) Prepare a request for a security waiver or exception.

    e. Maintain regulations, instructions, and directives.

        (1) Maintain regulations, instructions, and directives of higher authority pertinent to security and LE functions.

        (2) Develop supplemental guidance applicable to the activity.

    f. Approve CLEOC reports and records.

        (1) Approve IRs.

        (2) Approve criminal investigation reports.

        (3) Approve accident reports.

        (4) Approve PS surveys or inspections.

        (5) Approve access control records.

        (6) Approve desk journals.

  (7) Maintain field interview report files.

 g. Perform personnel administration.

  (1) Maintain security department personnel performance and training records.

  (2) Execute contract guard agreements.

  (3) Maintain oversight of reserve personnel performance and training records.

 h. Review plans for new construction or building modifications.

  (1) Coordinate with the public works officer and command security manager, review plans for construction of or modifications to a facility to ensure that all PS, force protection, and loss-prevention concerns are adequately addressed.

  (2) Make a recommendation for a security improvement to a building construction or modification plan.

  (3) Ensure compliance with the Unified Facilities Criteria (UFC).

 i. Implement/develop and maintain personnel identification and access control systems.

  (1) Inspect the personnel identification system for an activity or installation.

  (2) Inspect automated entry control systems (AECSs) for an activity or installation.

 j. Maintain working relationships.

  (1) Maintain frequent contact with the RSO for all regional policy and guidance.

  (2) Maintain contact with and solicit advice from the regional/installation staff judge advocate (SJA) concerning the legal aspects of LE and security.

  (3) Maintain a working relationship and agreements with the local Naval Criminal Investigative Service (NCIS) office.

  (4) Maintain working relationships with federal, state, and local LE agencies, other military activities, and host-nation (HN) officials.

  (5) Develop and maintain memoranda of agreements (MOAs) and memoranda of understandings (MOUs).

 k. Update security organization legal requirements.

  (1) Analyze a legal issue in conjunction with the SJA.

  (2) Change rules, regulations and procedures, and plans to reflect changes in law.

 l. Plan reserve security force use.

  (1) Plan reserve training.

  (2) Plan for integration and employment of reserve personnel into the NSF.

 m. Advise COs about LE capabilities.

 n. Prepare plans IAW Chapter 5 of this NTTP.

o. Establish installation policies IAW the commander's intent.

p. Provide input to the Navy Lessons Learned database.

q. Use the Defense Readiness Reporting System–Navy (DRRS-N).

3. Managing the following operational duties and shall be prepared to manage and perform any of the associated tasks listed below:

a. Direct crime prevention efforts.

    (1) Direct the execution of a key and lock control program.

    (2) Direct the completion of PS surveys for an installation.

    (3) Manage crime/loss-prevention programs.

    (4) Manage substance abuse education and program activities.

b. Direct confinement operations.

    (1) Manage security department compliance with detention cell or emergency plans and critical incident response plans when incarcerating a detainee or prisoner.

    (2) Manage the procedures used in transporting a prisoner.

c. Direct security efforts.

    (1) Advise tenant command security officers pertaining to protection efforts, including PS and LE.

    (2) Manage port security operations.

    (3) Manage installation or activity security incidents.

    (4) Manage compliance with post orders.

    (5) Manage compliance with personnel arming and weapons-carrying procedures for duty sections, including arms, ammunition, and explosives (AA&E) screening.

    (6) Manage vehicle inspection teams.

    (7) Evaluate and supervise manning of all security department functions and posts.

    (8) Coordinate the implementation of the security department response in support of the installation AT and EM plans.

    (9) Manage the performance of personal and vehicle identification, access control, and inspection at entry control points (ECPs).

    (10) Manage compliance with use-of-force procedures/ROE.

    (11) Supervise NSF at off-base military operations.

    (12) Coordinate the use of NSF at a high-risk incident.

(13) Conduct security briefings.

(14) Manage area search operations.

(15) Lead the security department response during civil disturbance operations.

(16) Manage security patrols.

(17) Lead the security department response during natural disasters.

(18) Supervise the establishment of on-scene incident command posts (ICPs).

(19) Participate in the functioning of the installation emergency operations center (EOC).

(20) Supervise the establishment of inner and outer security perimeters around the site of incidents.

(21) Supervise establishment of communications with a hostage taker or barricaded suspect until a responsible agency (NCIS/Federal Bureau of Investigation (FBI)) arrives on-scene.

(22) Supervise traffic control operations around incident sites.

(23) Supervise installation courtroom security.

d. Direct law enforcement efforts.

(1) Supervise NSF to ensure that actions are within assigned authority and jurisdiction.

(2) Supervise LE operations and incidents.

(3) Supervise performance of criminal complaint-handling procedures.

(4) Supervise the enforcement of U.S. customs laws for a unit departure to or return from a foreign country. At some overseas installations, this is not an NSF or security department function.

(5) Enforce compliance with the Posse Comitatus Act (18 U.S.C. 1385) during NSF operations off-installation or in response to a request for assistance by a civilian law enforcement agency.

(6) Apply applicable provisions of the Uniform Code of Military Justice (UCMJ)/Manual for Courts-Martial (MCM) to the processing of criminal incidents.

(7) Apply the Assimilative Crimes Act (18 U.S.C. 13) to the processing of criminal or traffic incidents.

(8) Ensure compliance with civilian authority cooperation policies.

e. Direct MWD and contract explosive detector dog (EDD) efforts.

(1) Coordinate MWD support in response to Navy, other Service, or other agency request.

(2) Supervise MWD and contract dog searches and inspections.

(3) Conduct security department and appoint disinterested-party inventories of MWD program training aids.

(4) Conduct inspections of MWD kennel facilities.

(5) Certify MWDs and contract EDDs in coordination with the CO.

    (6) Approve MWD rules, regulations, and procedures (RRP).

    (7) Designate in writing a primary and an alternate drug training aid and explosive training aid custodian.

    (8) Ensure only command-certified MWDs or contract EDDs are sent on EDD missions.

    (9) Approve MWD monthly reports.

  (10) Schedule, observe, and approve contract dog operations.

  (11) Ensure the health and humane treatment of assigned MWDs.

f. Direct investigative efforts.

    (1) Approve investigative inquiries.

    (2) Supervise evidence-handling procedures during LE investigations.

    (3) Supervise the use of an evidence custodial system.

    (4) Supervise the execution of search and seizure of evidence.

    (5) Appoint a primary and alternate evidence custodian.

    (6) Appoint a disinterested third party, senior to the primary and alternate custodian, to inventory the evidence locker.

    (7) Assign/supervise criminal investigations.

    (8) Approve criminal intelligence-gathering methods/sources.

    (9) Coordinate criminal investigations with NCIS.

  (10) Brief the commander on the results of criminal investigations.

  (11) Approve the use of cooperating witnesses during LE investigations.

  (12) Supervise crime scene examinations.

g. Recommend procedures for technical assistance to the CO on security and LE matters.

    (1) Use of nonlethal weapons (NLWs).

    (2) Use of MWD teams to identify a threat or evidence.

    (3) Use of NSF to resolve a threat.

    (4) Possible courses of action during an incident.

    (5) Acquisition of a PS system or equipment.

h. Direct traffic operations.

    (1) Provide recommendation(s) for changes to the installation traffic regulations and policies.

(2) Supervise traffic accident investigations.

(3) Review traffic accident investigation reports.

(4) Supervise traffic law/regulation enforcement operations aboard the installation.

(5) Recommend traffic control options to the CO during events or construction.

i. Direct safety procedures.

(1) Ensure compliance with Navy Tactical Reference Publication (NTRP) 3-07.2.2, Force Protection Weapons Handling Procedures and Guidelines. Publish weapons-clearing procedures.

(2) Inspect emergency vehicle safety equipment at security units.

(3) Supervise compliance with emergency response procedures during incidents.

j. Direct compliance with legal requirements.

(1) Supervise NSF compliance with legal restrictions and procedural requirements during LE operations.

(2) Conduct evidence inventories.

k. Manage assistance programs.

(1) Supervise NSF compliance with the requirements of the victim/witness assistance program (VWAP) when processing a victim or witness.

(2) Supervise NSF compliance with the requirements of the sexual assault prevention and response (SAPR) program when processing victims.

(3) Provide representation/assistance to the installation family advocacy program (FAP).

4. Performing the following personnel management duties and shall be prepared to perform any of the associated tasks listed below:

a. Organize the security force.

(1) Manage the security department table of allowance (TOA).

(2) Establish fixed and mobile guard posts and patrol zones for the installation.

(3) Review civilian security guard contracts.

(4) Coordinate with the administrative department to understand the activity manpower document (AMD), officer and enlisted distribution and verification report(s)/enlisted distribution and verification report(s)), and civilian placement reports.

b. Guide security force members.

(1) Assist NSF with career development.

(2) Supervise preparations for NSF temporary additional duty/temporary duty (TAD/TDY) missions.

(3) Inspect NSF to ensure that all uniform, grooming, and appearance standards are satisfied.

(4) Direct NSF members' compliance with orders, instructions, directives, regulations, rules, agreements, or procedures.

(5) Ensure compliance with the Lautenberg Domestic Violence Amendment to the Gun Control Act of 1968 (Title 18, U.S.C.).

(6) Manage the installation field training officer (FTO) program.

c. Maintain good order and discipline of the security department.

(1) Conduct internal affairs investigations.

(2) Interact with contract guard on-site supervisors relative to contract guard discipline.

5. Performing the following logistics duties and shall be prepared to perform any of the associated tasks listed below:

a. Prepare fiscal documents.

(1) Prepare security department budgets.

(2) Identify constraints in the application of resources.

(3) Prepare long-range budget plans.

(4) Publish operating target (OPTAR) budgets.

(5) Understand the defense Planning, Programming, Budgeting, and Execution System funding process.

b. Acquire assets.

(1) Recommend equipment purchases.

(2) Prepare funding requests.

(3) Approve purchases.

c. Maintain assets.

(1) Conduct AA&E and NLW inventories.

(2) Conduct inspections of organizational equipment.

(3) Conduct inspections of individual equipment.

(4) Supervise the use of administration procedures during an AA&E issue, return, or inventory.

(5) Supervise maintenance of uniforms, equipment, and vehicles.

6. Performing the following training duties and shall be prepared to perform any of the associated tasks listed below:

a. Identify training requirements.

(1) Identify individual security department personnel training requirements.

(2) Publish a long-range training plan.

(3) Publish an annual fiscal year training plan. (The approving security officer should sign and date the plan.)

(4) Publish an annual fiscal year exercise plan.

(5) Supervise the creation or maintenance of training records using the Enterprise Safety Application Management System.

(6) Authorize and document all changes to training schedules.

(7) Ensure all newly hired DON civilian police (GS-0083) and security guards (GS-0085) receive initial training at a Regional Training Academy or from a Regional Training Academy mobile training team.

(8) Ensure contract guards meet training standards as required by the statement of work (SOW).

b. Direct the performance of NSF training.

(1) Supervise training sessions.

(2) Evaluate prospective training curricula to ensure compliance with skill set requirements.

(3) Coordinate ASF training with tenant commands.

(4) Coordinate NSF school requirements and attendance of a security officer/MA/Navy civilian to a formal school.

(5) Confirm that the training staff has ready access to the Navy Knowledge Online (NKO) websites and other training materials.

c. Evaluate training efficiency.

(1) Inspect training records.

(2) Evaluate NSF members' mastery of required knowledge, skills, and abilities.

(3) Conduct required exercises to evaluate training effectiveness and lessons learned.

(4) Supervise FTOs in the performance of training new NSF members.

(5) Supervise NSF members' on-the-job performance.

(6) Ensure that NSF members complete the required reading, including use of force; all local post orders; rules, regulations, and procedures; crisis response plans; rules of engagement; and other documents as deemed appropriate.

(7) Ensure that vulnerabilities identified during training evolutions have been recorded in DRRS-N as a part of the command's self-assessment.

(8) Evaluate individual compliance with the training requirements listed in this NTTP.

7. Performing the following communications duties and shall be prepared to perform any of the associated tasks listed below:

   a. Disseminate information.

     (1) Prepare criminal statistics reports.

     (2) Supervise the input of information into the consolidated law enforcement operations center (CLEOC).

     (3) Present security or LE briefs.

     (4) Establish security department notification procedures.

     (5) Publish special orders.

   b. Manage security department communications assets.

     (1) Publish a communications plan, including considerations for and with local authorities.

     (2) Recommend communications equipment purchases or funding requests.

     (3) Inspect communications interoperability with emergency response organizations aboard the installation or region.

     (4) Review maintenance records for communications equipment.

     (5) Supervise communications procedures at emergency communications/dispatch centers.

     (6) Supervise communications procedures of NSF members.

     (7) Verify that emergency dispatchers have received the appropriate DoD Telecommunicator Level 1 Certification for Operators and Level II Certification for Supervisors. Additional certifications may be required if the emergency dispatcher is also responsible for fire and emergency services dispatch.

2.5 NAVY SECURITY FORCE LAW ENFORCEMENT DIVISIONS AND BRANCHES

2.5.1 Administration Division

The administration division provides oversight for all administrative and logistics functions determined to be not inherently governmental within the security department.

1. Maintain records and correspondence.

2. Retain and manage office security records as provided by reference SECNAV M-5210.1, Department of the Navy (DON) Records Management Manual, and SECNAV M-5210.2, Department of the Navy (DON) Standard Subject Identification Code (SSIC) Manual.

3. Maintain regulations, instructions, and directives.

4. Submit reports.

5. Perform personnel administration.

2.5.1.1 Supply/Logistics Branch

The supply/logistics branch is responsible for preparing budgets and fiscal documents and acquiring resources to operate and maintain the security department. This division executes the budget process and acquires and maintains assets. It also ensures vehicles, emergency and other equipment, arms, ammunition, and NLW are acquired and maintained and associated records are up-to-date.

2.5.1.2 Pass and Identification Branch

The pass and identification branch is responsible for preparing visitor passes at the visitor control center, verifying visitor qualifications for installation access, and issuing installation specific identification cards as directed.

2.5.2 Physical Security Division

The physical security division provides oversight for all PS functions within the security department. These functions can be either inherently governmental or not, depending on the issue. The division can be manned by a combination of military, civil service, and contractor personnel. It is responsible for formulating the policies, procedures, and measures to safeguard personnel, property, equipment, and facilities within the installation. PS programs will provide the methods to counter threats against the installation. PS measures are PS equipment, procedures, or devices used to protect security interests from possible threats. Primary functions within the physical security division include the following:

1. Review plans for new construction or building modifications to ensure that all PS, AT, force protection, and loss prevention concerns are adequately addressed.

2. Implement and maintain personnel identification and access control systems.

3. Conduct installation security assessments.

4. Prepare PS plans.

5. Conduct PS surveys

6. Conduct inspections of all critical areas.

7. Conduct annual vulnerability assessments.

8. Provide support to the AT working group.

9. Identify countermeasures to an identified vulnerability.

10. Identify the personnel, property, structures, and assets requiring protection aboard an installation.

The Navy's crime prevention branch is a specialized organization within the physical security division. This branch anticipates, recognizes, and appraises the crime risk and recommends action to remove or reduce it. Crime prevention is a proactive method to reduce criminal opportunities, protect potential victims, and prevent property loss.

1. The Navy crime prevention program is directed at the quality of life in the Navy. It presents military personnel, civilian employees, and dependents with programs designed to reduce their chances of becoming the victims of crime.

2. The designated crime prevention coordinator is responsible for coordinating crime prevention functions, including:

   a. Publish standard operating procedures (SOPs) for crime prevention.

b. Develop installation-specific crime prevention surveys.

c. Maintain an active liaison with NCIS, local LE agencies, and other organizations to provide for an exchange of crime prevention-related information.

d. Offer crime prevention services that include briefings and surveys to tenant organizations.

e. Participate in crime prevention campaigns to highlight specific crime problems for intensified public awareness efforts. When possible, Navy campaigns should coincide with national crime-prevention campaigns.

f. Ensure crime prevention practices are a part of the daily operations of the security department, involving all members of the department.

g. Use all available records and information to develop and maintain trends and analysis information.

h. Advise the SO, investigations and operations personnel, and NCIS on observed trends and analysis information.

3. Crime prevention surveys shall be conducted for living quarters (both family and bachelor and off-base housing where applicable); at nonappropriated fund/retail activities; and at Morale, Welfare and Recreational facilities on request. These surveys are designed to assess the vulnerability of the location as a target for crime and to provide recommendations to reduce the vulnerability. The security department should keep a record of all crime prevention surveys conducted and follow up to see if recommendations are implemented. Crime prevention records shall be maintained for a period of 3 years.

2.5.3 Operations Division

The operations division is responsible for all operational aspects of the security department at the installation, including daily supervision and oversight of the following divisions: patrol, harbor security boats, ECP, commercial and mobile vehicle inspection, MWD, and communications (where applicable).

2.5.3.1 Patrol Branch

The patrol branch maintains patrols to provide a protective presence to deter crime; responds to calls for assistance, alarms, and reports of criminal activity; enforces laws, regulations, and directives in areas of command jurisdiction; apprehends and processes subjects; enforces traffic laws and regulations; investigates traffic accidents; provides timely response to noncriminal service requests; and provides escorts.

2.5.3.2 Military Working Dog Branch

The military working dog branch is responsible for patrol, EDD, and drug detection dog (DDD) operations. The MWD branch should be fully integrated into the operations division and involved in LE operations. EDDs may be government or contract dogs. Patrol and DDDs are always MWDs. Working dogs will be used to enhance officer safety, detect explosives and other dangerous articles, and locate dangerous narcotics and controlled substances.

2.5.3.3 Guard Branch

The guard branch is responsible for NSF personnel that perform security guard functions within the installation. This includes the personnel administration within the security department for contracted guard services on an installation.

2.5.3.4 Animal Control Branch

The CO ensures that free-roaming domestic animals are collected as necessary. This function may be assigned to the NSF and training on capturing stray animals should be provided locally prior to performing this function. The installation veterinarian ensures that the installation confinement and disposition policy is published at least quarterly in an installation media source. SECNAVINST 6401.1 (series), Veterinary Health Services requires the CO to support and give command emphasis to programs that control privately owned and stray animals at large on military installations through capture, impoundment, disposal, or other physical means. The use of MWD kennels for privately owned or stray animals is prohibited.

2.5.3.5 Game Warden Branch

Title 16 U.S.C. 670, the Sikes Act, requires DOD organizations support State fish and wildlife agencies and carry out a program to provide for the conservation and rehabilitation of natural resources on military installations. The SO and installation environmental program managers shall prepare an integrated natural resources management plan that includes interoperability and cooperation with United States Fish and Wildlife Service, and the appropriate state fish and wildlife agency for the state in which the military installation is located. The resulting plan for the military installation shall reflect the mutual agreement of the parties concerning conservation, protection, and management of fish and wildlife resources. This may or may not involve the use of NSF for enforcement of fish and game laws. On installations with hunting, trapping and/or fishing that require enforcement support by NSF, the SO shall enter into MOUs/MOAs with state and/or federal enforcement agencies for adequate training or budget for trained NSF to enforce environmental protection programs.

2.5.4 Investigations Division

The investigations division provides investigative personnel to respond to crime scenes and assume responsibility or notify NCIS for assumption of major crimes; maintains working relationships between the NCIS and Navy commands/activities in the investigation and prosecution of crimes. Referral of investigations and regular communications are keys to maintaining good working relationships. Together with the NCIS representative that provides service to the installation, the SO shall establish a case review process.

1. The command investigator should seek to participate jointly with NCIS when NCIS has assumed investigative responsibility and when related to installation interests.

2. All cases investigated by NSF shall be documented with an IR.

3. Liaison shall be conducted with other military LE agencies and NCIS.

4. Where regionalization has been implemented, the investigative responsibility will fall upon the regional investigations coordinator (RIC), with coordination between the RIC and the RSO and SO. In those instances where conflict between this NTTP and the regionalization process exists, the RSO and RIC must submit the issue to the REGCOM for resolution.

2.5.5 Training Division

The training division is responsible for ensuring all personnel performing LE and PS duties have completed minimum training as required in OPNAVINST 5530.14 (series), Navy Physical Security and Law Enforcement Program. The training division shall also schedule all newly hired civilian police (GS-0083) and security guard (GS-0085) for initial training at a Regional Training Academy (RTA) prior to performing LE or PS duties. To ensure adequacy of training, each security department shall have an active FTO program. In addition, SOs or their designees shall review training records quarterly to ensure all personnel have received required training and immediately schedule personnel who are delinquent. Guard mount will be used to provide training and to ascertain NSF training knowledge. Deadly force and the use-of-force continuum shall be addressed at every guard mount. Additional training requirements, including crime prevention, community policing, and locally available training, should be explored to increase functionality of LE personnel. The training division shall also ensure that all NSF personnel who operate emergency vehicles have completed the Department of Transportation (DOT) Emergency Vehicle Operator Course (EVOC) from a certified instructor, at least once every three years.

1. Identify training requirements.

2. Conduct NSF sustainment training and ASF initial and sustainment training.

3. Evaluate training efficiency.

4. Managing security personnel qualification and certification programs to include mandates for the use of NSF PQS.

5. Managing the security drill and exercise program

6. Performing in organization exercise and assessment teams.

2.5.6 Communication Division

The communications division is subordinate to the Installation Emergency Management Officer but receives input from the Security Department Operations Division on operational functions. The communication division is responsible for radio, telephone, and computer functions, including operation for normal and routine transmission of data and dissemination of information, both voice and electronic. It maintains logs pertaining to tracking calls for assistance, including criminal and accidental in nature as well as patrol and fixed-post radio and telephone and e-mail transmissions.

2.6 ENVIRONMENT

The environment in which the law enforcement element operates is that of an overarching security department responsible for various facets of protection, which includes aspects of LE, PS, and AT on an installation and at off-station assets with exclusive jurisdiction such as housing. The environment will vary depending on geographic and diplomatic conditions but may include U.S. or foreign military installations, corrections/brig facilities, civilian shipyards, courtrooms, housing areas, fleet concentration areas, North Atlantic Treaty Organization (NATO) facilities, and forward operating bases as examples.

INTENTIONALLY BLANK

CHAPTER 3

Command and Control

3.1 OVERVIEW

Command and control (C2) is the process commanders use to plan, direct, coordinate, and control forces to ensure mission accomplishment. Communications systems used for C2 must be secure and redundant to prevent compromise of the commander's ability to execute C2. The National Incident Management System (NIMS) is the governing authority on incident control, and all LE personnel should be familiar with the basic tenets of incident control.

3.2 NOTIONAL COMMAND AND CONTROL STRUCTURE ASHORE FOR LAW ENFORCEMENT

3.2.1 Installation Commanding Officer

The CO shall perform and coordinate all Navy security program requirements within the installation's area of operations (AO). Commanders shall provide for initial and immediate response to any incident occurring on the installation. Additionally, commanders must contain damage, protect property and personnel, and restore order. To perform these functions, commanders may order searches, seizures, and apprehensions and take other reasonably necessary steps to maintain law and order.

3.2.2 Installation Security Officer

Through the WC, shift supervisor, or chief of the guard, the SO executes C2 of NSF responding to all LE incidents. The SO further coordinates with other emergency responders and serves as a tactical adviser to the CO. For major LE incidents under control of the CO, the security officer may perform the duties of the incident commander.

3.2.3 Command Duty Officer

As the CO's representative, the command duty officer (CDO) carries out the routine of the installation in the absence of the CO. The CDO works closely with the SO or designated representative in response to an LE incident as required, until such time as the CO or other designated officer in the chain of command assumes control.

3.2.4 Watch Commander

The WC is the designated SO representative and has authority of the SO when on duty. On major incidents the WC may be assigned as the incident commander or assistant incident commander.

3.2.5 Dispatcher

Dispatchers handle radio and other communication traffic and ensure the chain of command is properly notified as required by the severity of the incident and IAW the region/installation policy and the incident commander. Dispatchers must ensure pertinent communications are provided to all required personnel in a timely manner. Although not LE by assignment, dispatchers must be responsible to make decisions and dispatch appropriate personnel depending on the nature and severity of the LE incident IAW established protocol within the commander's intent.

3.2.6 Patrol Supervisor

The patrol supervisor is the roving supervisor for an area or a section of NSF personnel. This individual will usually be the first supervisor on site at an incident and makes decisions pertaining to the requirements of an LE incident and may assume the role of the incident commander on incidents requiring a larger response than one patrol member and/or one command investigator. The patrol supervisor does not automatically assume the duties of incident commander and may be advisory only on minor incidents.

3.2.7 Chief of the Guard

The chief of the guard is the senior NSF member of an assigned area, usually a pier, and performs the duties of the patrol supervisor of the assigned area.

3.2.8 First Responders

These are patrol officers or command investigators on scene at an LE incident. They perform all basic LE duties, including apprehending suspects, investigating crimes and accidents, and a myriad of tasks specified in Chapter 2 of this NTTP. For most LE incidents this first responder assumes the duties of the incident commander until relieved by a superior or the incident is concluded.

3.2.9 Fixed Posts

Personnel assigned to a fixed post may have LE authority but in this assignment are primarily AT or PS and should maintain situational awareness. If an LE incident occurs, a patrol supervisor or command investigator should respond to handle the incident. Criminal incidents may be a diversion to permit other activity to take place after distracting a watch. NSF at fixed posts will not be assigned as incident commander unless relieved of their fixed-post duties.

3.2.10 Incident Commander

The incident commander is in charge of the incident and the crime scene, makes decisions, and ensures appropriate action is taken based on the nature and severity of the incident. The incident commander is responsible for the successful resolution of the incident, including ensuring all aspects of evidence collection, reporting, cleanup, and a myriad of other tasks are completed. For serious incidents responsibilities may include managing multiple personnel and units as well as personnel from other agencies/units performing specific tasks within their area of expertise. The incident commander must have appropriate authority and rank to deal with personnel from other commands and agencies and the media. The incident commander must select a planned course of action, communicate the plan, and monitor execution of the plan while adapting as necessary IAW the Incident Command System (ICS).

3.2.11 Communications

All watch positions and functionalities addressed above must be supported by a communications architecture that enables them to receive information and provide direction to available assets as well as coordinate with nonorganic organizations. Enterprise land mobile radio (ELMR) will facilitate many of the requirements once rolled out and fully integrated. The following should be considered in the selection of the communications system(s) to be used:

1. Primary and secondary systems are required (redundancy).

2. Establish "talk groups" to be effective, and avoid "talk-over."

3. Networks should provide secure line of communications.

4. Encryption is required but does not need to be classified.

5. Systems must support mobile/all-weather operations.

6. Repeaters and amplifiers may be needed to reach remote locations.

7. Protocols, procedures, and capability must be provided to support communication/coordination with other government agencies. Therefore, plain language is the required method of communication.

8. Systems must comply with Federal Communications Commission (FCC)–designated frequencies.

9. Systems must be NIMS capable/compliant.

10. Training must be conducted to support proper radio/telephone/computer procedures, protocol, and discipline.

3.3 COMMAND AND CONTROL PROCESS

3.3.1 Fundamentals

The decision and execution cycle includes initiating a call for service. This can be accomplished by a self-initiated action when a member of the installation NSF observes an action that requires LE intervention, by a call received at dispatch requesting LE response, or by a person contacting a base NSF in person and alleging an action that may require LE action. Decision-making processes are used by the person receiving the request for service, to prioritize the response in order of potential of loss of life, severe bodily injury, and ongoing incident versus delayed reporting and to determine what level or number of LE personnel should be assigned.

Methods of control include communication processes to notify required LE forces to respond to the call for service. This could be by radio, telephone, email, or personal contact.

3.3.2 Recommended Unity of Command/Effort Principles

1. One command center for all security forces coordinates and directs all NSF, including incident commander, patrols, investigations, MWDs, other force protection and emergency management divisions, access control points, water-based, land-based, ship security, and decentralized control/decisionmaking.

2. Coordination and direction of all forces at any LE incident must be assigned to an incident commander. The incident commander may end up being a single patrol or a command investigator for a minor incident. Larger response incidents may require a hierarchy that may include the chief of the guard, the patrol supervisor, the WC, or the SO becoming the incident commander while the CO monitors the incident from a safe distance. When working some incidents, the incident commander may also be a member of fire, emergency medical services (EMS), or another unit based on the type of incident based on the ICS.

3. As incident complexity increases or decreases or focus changes, such as a fire being extinguished and arson suspected, there may be a need for the incident commander to be replaced, through an orderly process, by another person higher or lower in the chain of command or from another unit or agency. The change must be communicated to all involved personnel.

   a. Some major LE incidents may require establishing an ICP or EOC to free the incident commander and other personnel from routine duties during the incident. A decentralized ICP/EOC separates dispatch from the major incident so normal base LE operations may continue uninterrupted.

   b. The size and location of the ICP/EOC must take into consideration the location of the incident. The ICP/EOC must be close enough to effectively monitor and control the activities pertaining to the incident but remote enough that personnel reporting to and leaving the ICP/EOC do not interfere with activities at the scene or routine base operations.

   c. Whether mobile or a temporary fixed location, the ICP/EOC must be appropriately equipped to handle all required communications during an incident. The ICP/EOC could be as simple as a patrol member with a radio or as complex as a fully equipped and staffed trailer or building.

d. The size and capabilities of the ICP/EOC will increase and decrease IAW the scope and complexness of the incident.

e. The ICP/EOC shall be classified as a restricted area to ensure that only required personnel are present. The following are two examples of the Navy's "Way A-Head" state-of-the-art EOC:

(1) Regional operations center. The regional operations center is a NIMS-compliant multiagency coordination system using the incident/unified command system's organizational structure delineated in OPNAVINST 3440.17, Navy Installation Emergency Management (EM) Program, and FM 3-11.21/MCRP 3-37.2C/NTTP 3-11.24/AFTTP(1) 3-2.37, Multiservice Tactics, Techniques, and Procedures for Chemical, Biological, Radiological, and Nuclear Consequence Management Operations, to provide a collaboration point and operations center for regional staff to support execution of the regional EM plan; the regional AT plan; other supporting plans; defense support to civil authorities (DSCA) missions; the operational/contingency plans of assigned combatant, component, and fleet commanders; and the National Response Framework (NRF). The regional operations center will serve as the command, control, communications, computers, intelligence, surveillance, and reconnaissance point for an REGCOM to gather information, gain situational awareness, and exercise control over his forces across the entirety of the force protection and EM time lines, from early warning and detection of suspicious events through regional/installation response and recovery.

(2) Emergency operations center. The emergency operations center (EOC) is a NIMS-compliant multiagency coordination system using the incident/unified command system's organizational structure delineated in OPNAVINST 3440.17, Navy Installation Emergency Management (EM) Program, and Homeland Security Presidential Directive-5 (HSPD-5), Management of Domestic Incidents, to provide a collaboration point and operations center for installation staff to support execution of the installation EM plan; the installation AT plan; other supporting plans; DSCA missions; the operational/contingency plans of assigned combatant, component, and fleet commanders; and the NRF. The mission of the EOC is to support the incident commander or unified commander during emergencies with resource management support and establish strategic/operational-level objectives, as necessary. The EOC is responsible for coordination and liaison with local, other Service, and/or private response and recovery assets. From the EOC, the CO exercises operational control of installation forces and allocates resources.

4. The incident commander must understand the CO's intent. This is accomplished through training of all personnel who may be assigned duties that could result in their being an incident commander. This includes the purpose and any tasks that must be accomplished during an LE incident.

5. The incident commander must be permitted to use personal initiative within the guidelines of the NIMS.

6. The ratio of supervisors to subordinates in an incident should not exceed one supervisor to ten subordinates.

3.4 TRANSFER OF COMMAND AND CONTROL TO ANOTHER AGENCY

There may be instances where control of an incident will be transferred to another agency, such as NCIS, FBI, another federal agency, or a state or local civilian LE agency. The military forces employed in supporting the incident shall remain under military command and control at all times.

CHAPTER 4

Physical Security and Law Enforcement Principles

4.1 PHYSICAL SECURITY PROGRAM

Prevention and protection are the two primary concerns of the physical security program; both serve the security interests of people, equipment, and property. This NTTP establishes physical security as a supporting component of the protection function and describes defensive measures that enable protection tasks. The overall security posture at an activity includes policy and resources committed to safeguard personnel, protect property, and prevent losses. Physical security is further concerned with means and measures designed to achieve AT readiness.

4.2 PHYSICAL SECURITY PRINCIPLES

While the basic principles of physical security are enduring, security technology, components, and analytical tools continue to evolve and improve. Today, NSF have a full array of sophisticated systems, sensors, and devices to employ. The goal of the security system for an installation, area, facility, or asset is to employ security in depth to preclude or reduce the potential for sabotage, theft, trespass, terrorism, espionage, or other criminal activity. Listed below are the enduring principles of physical security:

1. Identify personnel authorized access.

2. Establish/define perimeters by some physical means, if only signage.

3. Establish defense-in-depth by creating layers of security before reaching assets.

4. Establish access control measures at the entry level to perimeters and/or enclaved areas.

5. Conduct surveillance detection to record the activities of persons behaving in a suspicious manner and to provide this information in a format usable by the appropriate law enforcement or intelligence officials.

6. Establish enclaves to provide a greater degree of protection to more critical access.

7. Establish screening process to prevent the unauthorized introduction of personnel or materials.

8. Identify threats and capabilities of those presenting potential threats.

9. Identify mission-critical assets and on-base critical infrastructure and provide these with a greater degree of protection.

10. Identify vulnerabilities that might allow a would-be threat to penetrate.

11. Mitigate vulnerabilities by using technology and/or concepts of operations to limit risk.

12. Participate in mass notification process to warn users/residents of potential threats.

13. Provide ability to prevent unauthorized entry.

14. Provide and maintain a response force capability.

15. Maintain comprehensive PS plans IAW Chapter 5 of this NTTP.

16. Exercise the plans regularly; develop lessons learned (observations and recommendations) and incorporate into subsequent PS plans.

17. Develop preplanned responses (PPR) through the use of SOPs to likely incidents.

18. Establish methods to mitigate and recover from criminal acts, natural disasters, or other harm and breaches.

19. Establish assessment methodologies to determine the impact of a threat.

20. Program for funding resources for measures necessary to protect an activity.

4.3 LAW ENFORCEMENT PRINCIPLES

The purpose of law enforcement is to prevent, interdict, and investigate crimes and to prosecute criminals. The prevention of criminal activity relies on a combination of security and deterrence. Security measures, ranging from locks and fences to surveillance cameras and patrols, make it more difficult to get away with crimes. Deterrence is in part a result of security and in part a result of the investigation and prosecution of crimes. Security and deterrence are not always easy to distinguish. NSF patrolling in cars conveys the impression that a crime might be observed and stopped outright; they also serve to remind the personnel they serve that NSF are present and are likely to find and prosecute the perpetrators of crimes. Listed below are the basic principles of law enforcement for Navy installations:

1. Investigate criminal activity, pursue violators, recover property, protect the military community, and maintain good order and discipline.

2. Enforce laws, through the use of police, to prevent crime and disorder.

3. Respond to calls for service.

4. Facilitate public awareness.

5. Seek and preserve public favor by constantly demonstrating absolute impartiality.

6. Direct NSF actions toward their functions and use their authority judiciously.

7. Using protective services efficiently in enforcing the law means the absence of crime and disorder, not the visible evidence of police action in dealing with them.

8. Use police intelligence to effectively manage scarce NSF resources.

CHAPTER 5

Physical Security and Law Enforcement Planning

5.1 OVERVIEW

Planning is imperative to ensure that PS and LE goals can be achieved when incidents occur, and to ensure that NSF personnel are informed of the commander's intent and can make decisions to secure the installation, deter criminal activity, investigate incidents, and apprehend those who commit criminal and other offenses aboard Navy installations. To avoid unnecessary duplication of effort and ensure unity of effort among host installations and tenant activities within the same geographical region, commanders will integrate the efforts of their security forces and develop joint plans with adjacent Navy assets or support agreements with other Services, local, state, and federal entities. It is especially critical that installations adjacent to special weapons areas plan, coordinate, and exercise security requirements jointly to provide for defense in depth. Legal advisers and SJAs should be consulted throughout the planning and execution phases to ensure compliance with U.S. domestic, international, and foreign law as applicable for the installation. A plan is not complete until it has been signed by the CO and exercised in its entirety.

Use the military decision-making process (MDMP) to provide a common framework for the installation staff to do parallel planning. It provides a logical sequence of decisions and interactions between the commander and the staff. The MDMP is a single established and proven analytical process to assist in developing estimates and plans. It helps the commander and staffs examine the situation and reach logical decisions. The seven steps of the MDMP are:

1. Receipt of mission

2. Mission analysis

3. Course-of-action development

4. Course-of-action analysis (wargaming)

5. Course-of-action comparison

6. Course-of-action approval

7. Production of orders or plans.

5.2 PHYSICAL SECURITY PLANNING

Physical security planning shall be integrated with LE and terrorist incident response planning. The commander should ensure that the command security posture is accurately assessed and security resources are appropriate to execute these programs IAW OPNAVINST 5530.14 (series), Navy Physical Security and Law Enforcement Program, and as established in Department of Defense Directive (DODD) 2000.12, Department of Defense Antiterrorism (AT) Program, and Department of Defense Instruction (DODI) 2000.16, Department of Defense Antiterrorism (AT) Standards. The commander shall prioritize assets and ensure that each is protected according to mission criticality, vulnerability, and the existing threat.

5.3 LAW ENFORCEMENT PLANNING

Each Navy installation should develop and publish LE RRP in agreement with regional guidance. Each SO shall also develop basic plans to respond to, investigate, and mitigate criminal actions that occur on the installation. Plans shall continually be evaluated on their effectiveness and modified as needed.

5.4 PLAN RESPONSIBILITIES

1. PS and LE plans shall incorporate requirements, policies, and procedures for facilities; equipment; all NSF, including regular, reserve, civilians, and ASF; employee training and education; and other aspects of LE-essential elements to ensure a facility is secure and maintain law, order, and good discipline on an installation.

2. LE planning shall be integrated with terrorist incident and EM planning.

3. LE planning shall include plans for all-hazard prevention and response.

4. Tenant and host commands shall ensure that intra/inter-service support agreements include complete and detailed LE requirements and responsibilities.

5. Integration and coordination of PS and LE planning and implementation of specific measures should be accomplished among Navy activities on a regional basis and with state, local, and HN LE.

6. Minimum required response to certain types of incidents must be addressed to protect NSF personnel.

5.5 PLANNING PROCESS

Every person and asset cannot be made totally safe; therefore, the objective of PS and LE planning should be to determine the level of protection and response appropriate for personnel and assets based on a threat analysis, with the goal of developing a realistic and cost-effective program. The planning process must identify what is to be protected and a list generated that identifies type, nature and criticality of assets. PS planning should include the following:

1. Using the threat assessment conducted by NCIS, installations will develop a list of most likely/dangerous threats from which they extrapolate a Design Basis Threat (DBT), from which the PS and AT plan to include resourcing strategy is based (see paragraph 5.8 of this chapter). This list shall identify those threats that will not be defended against and the rationale for this decision (risk management). Specific terrorist threat information may be classified in nature. If so this information will be handled in accordance with Navy classified material requirements.

2. Using electronic security systems (ESSs) and other technology to reduce both vulnerability to the threat and reliance on fixed security forces. ESSs are that part of PS concerned with the safeguarding of personnel and/or property by use of electronic systems. Systems include, but are not limited to, intrusion detection systems (IDSs), AECSs, and video assessment systems.

3. Integration of PS and AT measures into contingency, mobilization, and wartime plans, and testing of PS procedures and measures during the exercise of these plans.

4. Coordinating with installation operations security, crime prevention, information security, personnel security, communications security, EM, base engineers, public works, automated information security, and PS programs to provide an integrated and coherent effort.

5. Creating and sustaining security awareness.

6. Identifying resource requirements to apply adequate measures.

7. Include the installation antiterrorism working group (ATWG) in planning sessions.

LE planning should include the following:

1. Analyzing crime patterns to determine the most common offenses and where they are most likely to occur.

2. Integrating crime prevention and AT measures into contingency, mobilization, and wartime plans and testing procedures and measures during the exercise of these plans.

3. Training NSF in the installation Navy Mission-Essential Task List (NMETL).

4. Creating and sustaining crime prevention awareness.

5. Identifying resource requirements to support the plan.

5.6 PLANNING CONSIDERATIONS

The following factors should be considered in the determination of security and LE requirements and planning:

1. The terrorist and criminal threat as defined by NCIS, producing the installation's threat assessment for the AT plan

2. Ease of access to vital equipment and material

3. Location of high-crime areas

4. Location, size, and vulnerability of facilities within the activity and the number of personnel involved

5. Need for tailoring security measures to mission-critical operating constraints and other local considerations

6. Coordination of security and LE forces with Navy, federal, state, and local agencies/forces

7. Geographic location (existence of natural barriers and avenues of approach)

8. Legal jurisdiction within the AO

9. Mutual aid and assistance agreements

10. Local political climate

11. Adequacy of storage facilities for valuable or sensitive material

12. Possible losses and their impact on command mission and readiness

13. Possibility or probability of expansion, curtailment, or other changes in operations

14. Overall cost of security and LE operations and equipment

15. Availability of personnel and material

16. Analysis of Risk. An analysis must be conducted that identifies the risk a command is willing to accept, the justification, mitigation and approval by the CO.

5.7 CRISIS SITUATIONS

In evaluating security requirements, the possibility of injury to NSF must be considered, and LE and PS applications need to be applied to mitigate this threat. This issue is especially relevant when addressing measures taken during crisis situations (e.g., active shooter incidents, bomb threats, fires, terrorist incidents, hostage situations, or natural catastrophes). Incidents occurring within strategic weapons facilities shall be guided by sections pertaining to the protection of nuclear weapons in the Chairman of the Joint Chiefs of Staff Instruction (CJCSI) 3121.01 (series), Standing Rules of Engagement and Standing Rules for the Use of Force for U.S. Forces (U); SECNAVINST S8126.1, Navy Nuclear Weapons Security Policy; and SECNAVINST 5530.4 (series), Navy Security Force Employment and Operations. Some mitigation that offers protection to NSF should be written into the plans.

5.8 DESIGN-BASIS THREAT PROCESS

The design-basis threat is an estimate of the threat that faces installations across a range of undesirable events and based on the best intelligence information, intelligence community (IC) reports and assessments, and crime statistics available to the working group at the time of publication. However, users of the DBT must consider that undiscovered plots may exist, adversaries are always searching for new methods and tactics to overcome security measures, and the lone adversary remains largely unpredictable. The DBT establishes the characteristics of the threat environment to be used in conjunction with physical security standards. The intent of the DBT is to support the calculation of the threat, vulnerability, and consequence to an installation or facility when calculating risk and to determine specific adversary characteristics that performance standards and countermeasures are designed to overcome. The results of the DBT process will also assist in establishing a list of those threats that will not be defended against and the rationale for this decision (risk management).

5.8.1 Risk Assessments

1. The DBT provides specific details as to the characteristics of each event that might take place at an installation or facility. They are based on a worst-reasonable-case. Each event provides sufficient information from which the threat, consequences, and vulnerability can be estimated in the conduct of a risk assessment:

    a. A baseline threat rating is provided, and target attractiveness characteristics that may make a facility more attractive as a target (increase the threat) are enumerated as appropriate. These factors should be considered in determining a score or rating for threat. Deviation from this threat level should be fully documented and supported with current intelligence information.

    b. The specifics about the size, number, equipment, etc. included in the scenario can be used to estimate the potential consequences. Consequence estimates should be based on the potential effects of a successful undesirable event.

    c. The specifics of the scenario should also be used to measure the effectiveness of existing protective measures in determining vulnerability. The vulnerability score should reflect the likelihood of the existing countermeasure successfully resisting or overcoming the DBT event scenario.

2. Where appropriate, modifications to the event scenarios are permitted. However, modifications must be supported with a detailed rationale and should provide sufficient detail to support the quantification of threat, consequence, and vulnerability. Additionally, in estimating the threat level, specific information unique to the installation or facility may be used. Local crime statistics, the tactics of adversary groups known to be operating in a particular area, and other actionable intelligence that suggests a different threat level may modify the threat from the baseline. When used, this information must be fully documented.

5.8.2 Performance Standards

In designing countermeasures to defeat or mitigate specific events, the characteristics of the DBT event scenarios should be considered as design parameters for performance of a countermeasure. For example, when it is necessary to protect against a vehicle-borne improvised explosive device (VBIED), the device size specified for VBIED events should be used for engineering calculations.

5.8.3 Baseline Threat

1. An estimate of the relative threat posed to Navy installations will be provided by NCIS. Ratings include: VERY LOW, LOW, MODERATE, HIGH, or VERY HIGH. The baseline threat levels are estimates based on an analysis of available threat information. The baseline threat level is intended to address Navy facilities generically. When being used to develop national or broad standards, the baseline threat level is applicable.

2. When being used to determine threat scores for a specific installation, users are expected to tailor the threat level based on known threat information applicable to the specific location such as local crime rates, historical events at the installation, or known adversary organizations operating in the vicinity. Absent information that merits a modification, the baseline threat level is applicable.

5.8.4 Analytical Basis and Target Attractiveness

1. Historical incidents and trends of adversarial activities serve as the analytical basis for the DBT scenario baseline threat. The DBT baseline threat takes into account such things as target attractiveness considerations, provides the historical basis and capability of adversaries, and shows possible variations on the DBT scenario. The analytical basis section is by no means all-inclusive of the information that was considered in estimating the threat. Rather, it is a synopsis of the most relevant information.

2. Target attractiveness considerations are provided to aid in identifying aspects of a particular installation that may make it more or less likely to be a target of a particular undesirable event, subsequently modifying the baseline threat to the installation. Additionally, when applicable, factors are identified that may change the parameters of the event, such as indicating potential use of a larger explosive device, more adversaries, or more complex methods of attack. In assessing a specific installation, users are expected to consider target attractiveness factors and modify the threat rating as appropriate. An installation that embodies greater target attractiveness may face a higher threat. Whenever the threat level is determined to deviate from the baseline, the target attractiveness factors that influenced the rating must be documented and fully supported by detailed information as part of the assessment.

5.8.5 Undesirable Events

Figure 5-1 lists some undesirable events that could have an adverse impact on the operation of an installation, facility, or mission of the Navy.

Undesirable Event	Description
Aircraft as a Weapon	Attack on an installation using an aircraft as an improvised explosive device.
Arson	Accessing a facility and deliberately setting fire to the facility or to assets within the facility.
Assault	Physically assaulting (with or without a weapon) a person or persons on the installation or in a facility.
Ballistic Attack–Active Shooter	An active shooter is an individual actively engaged in killing or attempting to kill people in a confined and populated area, typically through the use of firearms.
Ballistic Attack–Small Arms	Firearm fired from off base into an installation or facility.
Ballistic Attack–Standoff Weapons	Mortar, rocket-propelled grenade, etc., fired from offsite into an installation or facility.
Breach of Access Control Point–Covert	Use of deceit, coercion, or social engineering to gain access to a facility through a controlled entrance.
Breach of Access Control Point–Overt	The use of force and/or weapons to defeat a personnel screening or access control checkpoint (including ID checks).
Chemical/Biological/Radiological (CBR) Release–External	Intentional release of a CBR agent into an installation or facility through a specific access point, such as air intake, windows, or doorways, from outside the facility.
CBR Release–Internal	Intentional release of a CBR agent carried into a facility, including in general interior spaces (lobbies) or into specific rooms or systems (HVAC rooms).
CBR Release–Mailed or Delivered	A CBR substance or dispersal device sent to the facility through U.S. Mail or a commercial delivery service, including an unwitting courier.
CBR Release–Water Supply	Intentional release of a CBR agent into an installation's potable water supply from a location outside the facility.
Civil Disturbance	Deliberate and planned acts of violence and destruction stemming from organized demonstrations on or near Navy property.
Coordinated or Sequential Attack	A planned assault on an installation that integrates the aspects of several undesirable events.
Disruption of Building & Security Systems	Physically accessing building or security systems for the purposes of disruption or manipulation of the systems.
Explosive Device–Mailed or Delivered	An explosive device sent to the installation through U.S. Mail or a commercial delivery service, including an unwitting courier.
Explosive Device–Man-Portable External	An explosive device placed on the property, outside of a building and left to detonate after the adversary departs.

Figure 5-1. List of Undesirable Events (Sheet 1 of 2)

Undesirable Event	Description
Explosive Device–Man-Portable Internal	An explosive device carried onto the installation by an adversary or an unsuspecting occupant, visitor, or courier, and left to detonate after the adversary departs.
Explosive Device–Suicide/Homicide Bomber	An explosive device carried onto the installation by an adversary with the intent of reaching a specific target or area then detonating, killing, or injuring the bomber and others.
Explosive Device–VBIED	An attack against an installation or facility that utilizes a vehicle to deliver an improvised explosive device.
Hostile Surveillance	The surveillance of key assets, personnel, security features, operations, or sensitive areas from offsite, or outside secure areas for the purposes of collection of information in preparation for an attack.
Insider Threat	Individuals with the access and/or inside knowledge of an organization that would allow them to exploit the vulnerabilities of that entity's security, systems, services, products, or facilities with the intent to cause harm.
Kidnapping	Abduction of an occupant or visitor from a facility, including from inside secured areas (e.g., a child care center) or outside on the installation (e.g., a parking lot).
Release of Onsite Hazardous Materials	Unauthorized access to hazardous materials stored on base with the intent of harming personnel or damaging facilities.

Figure 5-1. List of Undesirable Events (Sheet 2 of 2)

5.8.6 Design-Basis Threat Scenario Analysis

Figure 5-2 is an example of an analysis of a DBT regarding an overt attempt to breach an access control point. This is an example that can be used for any undesirable event with adequate analysis.

5.9 ACTIVE SHOOTER RESPONSE

This section provides recommendations for actions and tactics to be used when confronting an active shooter situation. Some of the more tragic events in our country's recent history have been episodes where a deviant has carried out shootings in public places. These shootings take place for no other reason than to harm as many innocent people as possible. They are often unpredictable and strike in places such as our schools, places of worship, and places of work. They also occur in random public settings. The definition of an active shooter incident is when one or more subjects participate in a shooting spree, random or systematic, with intent to continuously harm others. The "active shooter" is not a new phenomenon; however the frequency of these events seems to be on a rise. The active shooter's overriding object/goal is that of murder, often mass murder, rather than other criminal conduct, such as robbery or hostage taking.

5.9.1 Purpose

The term "active shooter" within this NTTP will also include anyone who uses any other deadly weapon (knife, club, bow and arrow, explosives, etc.) to systematically or randomly inflict death or great bodily harm on others. Active shooter scenarios are incredibly dangerous and difficult because there is no criminal objective (robbery, hostage-taking) involved other than mass murder. Often, the shooter has no regard for their own life, and may be planning to die. These elements leave NSF no other tactic than to find and neutralize the shooter as quickly as possible. For NSF, the key to protecting the public from an active shooter is to respond appropriately to the unique situation with intensity and speed.

Undesirable Event		Breach of Access Control Point–Overt			
Definition		The use of force and/or weapons to defeat a personnel screening or access control checkpoint (including ID checks).			
Original Assessment	02-28-11	Revision	1	Date	07-15-11
Classified Annex	No	Classification		Date	

Design-Basis Threat Scenario

An adversary uses a handgun in an effort to breach security at access control points with the intent to enter the installation.

Baseline Threat

Based on the unsophisticated nature and historic frequency of this type of event, the baseline threat to (name of installation) from this event is assessed to be **MODERATE.**

Analytical Basis

The presence of an access control point at the entrance to (name of installation) does not necessarily deter adversaries from attempting to forcibly penetrate the installation access control point. Between March 2008 and March 2010, lone adversaries were responsible for eight of thirteen violent incidents perpetrated or directed at government facilities, some of which involved shootings at access control points. Five of those shootings at government facilities killed 18 (including two of the shooters), and wounded 37 (including two of the shooters). Comparatively, between February 1993 and February 2008, lone adversaries were responsible for 5 of 15 violent incidents that were also perpetrated or directed at Federal government facilities, including three shootings. An example of this type incident is on 4 Mar 2010 when a 36-year-old gunman shot at close range and wounded two Pentagon police officers manning an access control point outside the Pentagon Metro station before they mortally wounded him as he attempted to run toward a Pentagon entrance. The shooter, armed with a handgun, previously had used podcasts and a Web page to document his anger against the U.S. Government.

Target Attractiveness

Lone adversaries were responsible for the majority of known attempts to forcibly breach access control points. The unpredictable nature of the motivations of lone adversaries makes it difficult to determine what specific factors will make a facility or individual a more attractive target to these individuals. (Name of installation) which houses multiple Flag Officers faces a higher threat of this event. Also, the presence of high value assets, material, information, etc. indicates a higher threat from this type of event.

Outlook

Based on recent violent attacks involving the breach of security, increased security measures and the vigilance of security personnel are vital in preventing future attempts by armed adversaries from breaching access control points at (name of installation).

Figure 5-2. Examples of an Analysis of a Design-Basis Threat

5.9.2 Definitions

1. Contact team. The first responding officers/security personnel, minimum suggested size of two and up to four, who pursue the active shooter. The focus and mind-set are to make contact as soon as possible and neutralize the threat.

2. Rescue team. The second set of additional NSF personnel arriving on-scene, who form a team to locate and rescue injured persons.

3. Ad hoc commander. Lead NSF temporarily in charge of the incident.

5.9.3 On-Scene Command

While it is important to provide medical treatment to the wounded or injured, the primary duty of NSF responding is to protect innocent life by stopping the actions of an active shooter(s). Regardless of rank, the first supervisory NSF who is not part of the contact or rescue team to arrive on the scene of any active shooter shall become the on-scene commander and remains in that capacity until properly relieved. The on-scene commander shall do the following:

1. Establishes an ICP.

2. Determines response and staging area for arriving personnel.

3. Requests additional resources, e.g., NSF response personnel, NCIS, local agency(ies)' special weapons and tactics team, or negotiator(s).

4. Arranges a safe-staging area for medical units and triage area.

5. If suspect is apprehended or incapacitated, ensures the crime scene stays secure and maintains integrity until investigators or NCIS arrives.

5.9.4 Stages of an Active Shooter

An active shooter is usually not a spur-of-the-moment actor. The person progresses through a number of identifiable stages. These stages may occur in rapid succession or over a period of months or even years. During the first four stages, security forces may have an opportunity to intervene before the shooter is able to execute the plan. These stages are:

1. Stage 1: Fantasy. Shooters usually begin the process by imagining the event. They may romanticize the media coverage and/or the notoriety attributed to them because of the event. Often at this stage, the shooter has become radicalized as a result of outside sources. Shooters may express these fantasies through web postings, writings, artwork, or even discussions of the event.

2. Stage 2: Planning. The next stage is to select a target. Decisions on who, what, when, and where are made during this stage. Shooters select their weapons of choice and determine the logistics of traveling to the site, transporting weapons, and other details of the event.

3. Stage 3: Preparation. During this stage, shooters obtain weapons and supplies necessary to carry out the planned event. Shooters may preposition these items prior to the target date. Shooters may also warn certain individuals to avoid the target location during a given time period.

4. Stage 4: Approach. At this point, shooters have developed a plan and obtained the necessary weapons and supplies, and are acting on the plans. Shooters are traveling to the target location and, most likely, are armed with their weapons of choice. Security forces may encounter shooters through a traffic stop, an ECP vehicle inspection, or a citizen complaint. Contact with shooters at this point is very dangerous. However,

by approaching in a tactically sound manner, security forces may be able to stop shooters before anyone is harmed.

5. Stage 5: Implementation. Shooters are executing the plan. Because they are highly focused on their targets, they will not stop until they run out of ammunition or victims or take their own lives. Responding NSF should rely on their training and take immediate appropriate action.

5.9.5 Initial Response

The first responding security officer to an active shooter event must recognize the real dangers inherent in armed or physical confrontations. Dispatch may have provided some basic information of the situation as the officer approaches, but that information may not be entirely accurate. Information that can be gleaned quickly from eyewitnesses may provide valuable insights. The following are important items to determine:

1. The number of armed assailants

2. The manner in which they are armed

3. A brief description of the suspect(s)

4. Where the suspects are located and the direction they are heading

5. Size of backup forces and arrival times.

The situation demands that NSF actively seek to engage and neutralize the threat. Never hesitate to use deadly force when necessary. Aggression is used to implement violent action to intimidate and overwhelm the active shooter. Actions must be IAW the RUF.

5.9.6 Active Shooter Response Fundamentals

5.9.6.1 Considerations for Tactical Movement

Because the security force must move quickly to locate and neutralize the active shooter, safe movement is imperative. Tactical movement provides cover, concealment, and readiness for any eventuality. Movement toward the suspect should make full use of all available cover and concealment areas. Tactical movement places an individual in a better position in relation to the adversary. This may mean protection from the adversary or movement to neutralize the threat. When moving in unison with a partner or other members of the contact team, personnel may employ the leap-frog technique for the safety of all individuals. The leap-frog maneuver is a bounding technique dictated by the speed of the individuals and the threat level. This technique allows personnel to provide fire cover for each other during movement, allowing them to move faster with greater security. For more complete tactics and techniques regarding tactical movements, refer to NTTP 3-07.2.1, Antiterrorism.

5.9.6.2 Actions While Moving

While trying to locate the threat using tactical movement, the security force must exhibit certain characteristics and perform several other simultaneous actions to enhance the chances of a swift and successful outcome. The security force must:

1. Become tactical decision makers. They must be assertive and decisive while remaining flexible to adjust to changing situations.

2. Maintain situational awareness while constantly assessing the evolving threat.

3. Communicate with and, when necessary, direct fellow security personnel and other responders, including medical and possibly fire personnel. Effective communication is vital to the success of the response.

5.9.6.3 Principles of Close-Quarters Battle

A high probability exists that the active shooter will be inside a structure and close-quarters battle (CQB) techniques will need to be used to locate and engage the threat. The principles of CQB include detailed planning, speed, and aggressive action. CQB is a tactical entry procedure used in confined spaces. These techniques are in direct contrast to the longer, open-range areas associated with other forms of engagement. An active shooter scenario can quickly turn into a hostage situation, requiring a highly trained specialized team to be called. If the active shooter continues to kill people, then the security force has no alternative and must attempt to neutralize the threat. Information regarding CQB can be found in NTTP 3-07.2.1, Antiterrorism.

5.9.7 Tactical Withdrawal

When an overwhelming resistance situation arises, the contact team should maintain a position that keeps the active shooter pinned in one location until reinforcements arrive.

5.10 RESPONSE PLANNING STANDARDS

Installations shall be provided an LE response capability. This may be provided by a nearby installation if the response capability can meet response standards as established in OPNAVINST 5530.14 (series), Navy Physical Security and Law Enforcement Program. Patrols should be able to respond to calls for service within 15 minutes. Emergent life-threatening calls shall be responded to as soon as possible with due regard for safety and traffic conditions.

5.11 INSTALLATION SECURITY OFFICER PLANNING RESPONSIBILITIES

While the SO should not be tasked with producing the EM plan, the SO should provide input into the planning process and be intimately familiar with the plan. NSF preplanned responses for EM incidents will be incorporated into EM plans vice creating independent stand-alone RRP. The EM plan shall include a continuity of operations plan that identifies alternate communication and arming point locations for the security department.

SOs shall develop preparation and response plans for all foreseeable hazards, including natural and man-made disasters such as hurricanes, earthquakes, forest fires, floods, leaks at chemical plants, oil spills, radiological contamination, and power outages; nuclear, biological, or chemical attack; or sabotage, including attacks against critical infrastructure as applicable for their installation. SOs shall ensure that the protection of the NSF is included as part of the plan IAW OPNAVINST 3440.17, Navy Installation Emergency Management (EM) Program. The SO and PS/LE staffs are responsible for performing the following planning duties and will be prepared to perform any of the associated planning tasks listed below:

1. Plan integrated security programs.

2. Conduct risk assessments.

3. Assess the installation LE program.

4. Assess the installation PS program.

5. Assess the installation crime/loss prevention program.

6. Assess the AT plan.

7. Identify countermeasures to mitigate security vulnerability.

8. Identify the personnel, property, structures, and assets requiring protection aboard an installation.

9. Identify and recommend resources necessary to implement an effective LE program.

10. Recommend locations and hours of operation of installation ECPs based on reliable data collected.

11. Recommend use of mobile inspection teams.

12. Recommend placement of barriers for the installation.

13. Recommend establishment of critical assets and restricted areas aboard the installation.

14. Identify and prioritize property susceptible to theft and pilferage.

15. Perform loss trend analysis aboard the installation.

16. Identify inconsistencies and recommend corrective action between mutually supportive plans.

17. Provide advice on correcting deficiencies found during security inspections or vulnerability assessments.

18. Publish a set of metrics to support mission goals and objectives of the NSF.

19. Identify fixed and mobile security and LE posts.

20. Direct weapons placement and fields of fire for the installation or ships in the waterside restricted area.

21. Conduct NSF operational planning.

    a. Plan LE operations.

    b. Plan investigative surveillance operations.

    c. Plan protective services operations.

    d. Plan investigative operations.

    e. Plan surveillance operations.

    f. Plan counterdrug operations in conjunction with NCIS.

    g. Plan shore patrol operations.

22. Coordinate with the training department to exercise plans and develop lists of lessons learned.

23. Revise plans using lessons learned from exercising plans.

5.12 INSTALLATION PLANNING CONSIDERATIONS REGARDING ARMS, AMMUNITION, AND EXPLOSIVES

SOs shall take AA&E into consideration when planning for overall installation security. Installation LE and PS plans should contain a listing of all areas aboard the installation in which AA&E are stored, the PS measures in place to protect those assets, what NSF response requirements are generated by ESSs, and contingency plans that include additional security for AA&E during periods of special vulnerability. More in-depth guidance regarding AA&E can be found in OPNAVINST 5530.13 (series), Department of the Navy Physical Security Instruction for Conventional Arms, Ammunition, and Explosives (AA&E).

In addition to planning for PS measures, security officers should consider AA&E storage areas as potential force multipliers for both potential adversaries and the NSF. Security planning should include response to potentially compromised AA&E storage facilities, as well as possible use of arms and ammunition assets stored in armories outside the security unit ready-for-issue room, should security unit spaces become unattainable.

5.13 NAVY SECURITY FORCE EMERGENCY AUGMENTATION PLANNING

As part of crisis management planning, guidelines shall be prepared for NSF personnel to provide additional security during emergencies. Plans shall provide for possible augmentation by the ASF, shipboard security reaction force (SRF), Navy Expeditionary Combat Command (NECC), or Marine Corps security forces (MCSF) and other additional personnel and equipment. When plans include local LE or other agencies, MOUs must be executed as part of the planning process. These plans shall also provide for the essential training of and coordination with augmentation personnel, as well as rapid identification and acquisition of emergency equipment and supplies. Procedures for requesting augmentation from assets not part of a region must be coordinated through the respective geographic combatant commander (GCC).

5.14 ADDITIONAL PLANNING RESPONSIBILITIES

Physical security and crime prevention coordinators' duties include:

1. Performing evaluations of physical protection programs for facilities and installations
2. Evaluating security procedures and protection systems
3. Reviewing facility design
4. Recommending security countermeasures to protect against vulnerabilities
5. Conducting security surveys and inspections
6. Evaluating NSF assets
7. Determining overall effectiveness of security programs
8. Preparing for and evaluating security exercises and training
9. Reviewing AT plans and programs
10. Evaluating AA&E security
11. Assisting in the development of AT plans and procedures
12. Assisting in the acquisition and evaluation of security systems.

5.15 APPLICABILITY TO ANTITERRORISM PLANNING

As all terrorist acts have elements of criminal behavior involved, separate PS and LE plans should be annexes to installation AT plans.

5.16 SUICIDE PREVENTION PLANNING

Suicide prevention is a commander's program. Support of the Sailor resides with all personnel across the enterprise, but ownership of the incident (metrics and data) should not be delegated to NSF. It is within the organization that the first signs, actions, or manifestations will occur. The NSF role and protocol is one of situation control (usually apprehension) or post-attempt action response, whether the attempt was successful or not.

Suicide is a preventable personnel loss that impacts unit readiness, morale and mission effectiveness. Substance abuse, financial problems, legal problems, and mental health problems (such as depression) can interfere with individual efficiency and unit effectiveness and also increase a person's suicide risk. NSF must be trained to recognize the symptoms displayed by persons at risk of suicide and understand the procedures to follow that decreases the suicide risk and ensures treatment of an at-risk individual as soon as possible. Suicides and suicide-related behaviors shall be reported to the installation commander. Following are definitions for various suicide-related behaviors.

1. A suicide is a self-inflicted death with evidence (either implicit or explicit) of intent to die.

2. A suicide attempt is a self-inflicted potentially injurious behavior with a nonfatal outcome that may or may not result in injury and for which there is evidence (either implicit or explicit) of intent to die.

5.16.1 Elements of a Suicide Prevention Program

Navy suicide prevention programs consist of four elements:

1. Training—increasing awareness of suicide concerns, improving wellness and ensuring personnel know how to intervene when someone needs help.

2. Intervention—ensuring timely access to needed services and having a plan of action for crisis response.

3. Response—assisting families, units, and service members affected by suicide behaviors.

4. Reporting—reporting incidents of suicide and suicide-related behaviors.

5.16.2 Suicide Prevention Program Training

Suicide prevention training shall be conducted at least annually for all NSF. General military training materials may fulfill part of this training requirement but must be supplemented with information on local action plans and support resources. Suicide prevention training should include, but is not limited to:

1. Protocols for responding to crisis situations involving those who may be at high risk for suicide; and contact information for local support services.

2. Recognition of specific risk factors for suicide.

3. Identification of signs and symptoms of mental health concerns and operational stress.

4. NSF duty to obtain assistance for others in the event of suicidal threats or behaviors.

5.16.3 Navy Security Force Response Procedures

1. Security departments shall develop written procedures to ensure NSFs routinely responsible for installation emergency response executes their suicide prevention program responsibilities. Additionally, all installation NSF shall receive annual training which reviews safety precautions, procedures, and de-escalation techniques when responding to situations of potential suicide-related behaviors and psychiatric emergencies. Procedures will be coordinated with other first responder organizations (fire, medical) to ensure consistency of response. Included must be NSF procedures when a person at risk for suicide is being transported by medical or unit personnel to a medical treatment facility.

2. If NSFs are in contact with a person who makes comments, written communication, or behaviors that indicate there is imminent risk that the person may cause harm to self or others, safety measures shall be taken that include restricting access of at-risk personnel to means that can be used to inflict harm. These procedures will include a search of the person to determine if the individual is in possession of a weapon

that could harm the person or the responding NSF. All evidence of suicidal intent should be retained as evidence. Additionally, command leadership must be immediately notified.

3. In the absence of guidance from a mental health professional, responding NSFs should recommend the following actions be taken by unit leadership until the person can be transferred to the care of a mental health professional:

   a. Keep the individual under continual observation or physical control until presented to proper medical authorities.

   b. Remove personal hazards (no weapons, belt, shoes, boot straps, draw strings, shirt stays, and personal hygiene items such as toothbrush or razor).

   c. Prevent access to environmental hazards (room free of sheets, elastic bands, mirrors, pencils, pens, window dressings (such as blinds), shoelaces, strings, alcohol, weapons, medication, cleaning supplies, razors, metal eating utensils, telephones, tools, rope, or any other breakable, or sharp-edged object).

INTENTIONALLY BLANK

APPENDIX A

Police Intelligence Operations

A.1 POLICE INTELLIGENCE OPERATIONS

Police intelligence operations (PIO) supports, enhances, and contributes to the commander's protection program and situational understanding by analyzing, integrating, and portraying relevant criminal threat and police information and intelligence that may affect the operational environment (OE). This threat information is gathered by NSF as they conduct security functions and by other Navy personnel, other Services, local LE organizations and host nation security forces.

PIO is a Navy security forces police function, integrated within all security department operations, that supports the operations process through analysis, production, and dissemination of information collected as a result of police activities to enhance situational understanding, protection, civil control, and law enforcement. Installation security departments do not normally have dedicated PIO analysts but instead must rely on the ability of assigned NSF to perform this function. This function is normally assigned to the Physical Security or Investigations Division. Information gathered as a result of PIO (whether information directly supporting LE investigations or IR generated by a commander or staff) is gathered while conducting NSF police operations. Upon analysis, this information may contribute to a commander's critical information requirement (CCIR) and focus policing activities required to anticipate and preempt crime or related disruptive activities to maintain order. Other key definitions that provide framework and understanding for police intelligence include the following:

1. Police information is all available information concerning known and potential enemy and criminal threats and vulnerabilities collected during police activities, operations, and investigations. Analysis of police information produces police intelligence.

2. Police intelligence results from the application of systems, technologies, and processes that analyze applicable data and information necessary for situational understanding and focusing policing activities to achieve social order.

3. Criminal intelligence is a category of police intelligence derived from the collection, analysis, and interpretation of all available information concerning known and potential criminal threats and vulnerabilities of supported organizations.

NCIS and Installation Security Department staffs provide criminal intelligence analysis to commanders that identify indicators of potential crimes and criminal threats against Navy property, facilities, and/or personnel. Criminal intelligence is a subset of police intelligence focused on criminal activity and specific criminal threats. It is more focused in scope than police intelligence, which has a broader focus that includes police systems, capabilities, infrastructure, criminal activity, and threats. All criminal intelligence is police intelligence; however, not all police intelligence is criminal intelligence.

PIO provides essential products and services in support of security operations at Navy installations. In particular, police intelligence provides the commander and security officer with situational understanding necessary to reduce threats against Navy installations; provide threat intelligence for in-transit security; and focus the development and implementation of threat countermeasures to safeguard Navy personnel, material, and information. Regardless of the OE, PIO helps bridge the information gap between what a commander knows and does not know. PIO provides direct support to the intelligence cycle, with the most reliable information being obtained through effective PIO networks.

A.1.1 Bridging the Information Gap

PIO activities provide capabilities that can bridge the gap between traditional military intelligence and information focused on policing and the criminal environment. When the intelligence staff officer identifies a gap in the commander's knowledge of the threat and the current threat situation, that gap may be included as PIR or selected as indications and warnings (I&W). The intelligence officer will then develop a collection plan to help the commander in filling this information gap. Installation security staffs will also identify information gaps pertinent to policing activities and develop information requirements (IR) to fill those gaps. The IR findings may also be included in the PIR of the echelon commander.

Note

> Intelligence personnel are restricted from collecting information on U.S. persons (see Department of Defense [DOD] 5240.1-R) and will not direct security personnel to conduct such DOD collection activities.

Part of the commander's collection strategy is to select the best collection asset available to cover each information requirement. After a thorough analysis—an analysis that includes availability, capability, and performance history—a decision is made as to which collection assets can best be used. In security departments, the SO will identify information gaps; develop IR; develop, synchronize, and integrate collection plans; and task or coordinate for collection assets.

In the United States or its territories, effective PIO can provide installation commanders with situational understanding and address information gaps to ensure that threat assessments are valid and reliable. PIO in support of installations is conducted by the SO and their staffs responsible for LE operations. NCIS elements conduct PIO in their AO to support investigative and other support requirements. PIO capitalizes on connectivity between installation LE and civilian domestic agencies. In this environment, Navy intelligence involvement is typically limited. According to Department of Defense Directive (DODD) 5200.27, excluding specific LE and protection-related missions "where collection activities are authorized to meet an essential requirement for information, maximum reliance shall be placed on domestic civilian investigative agencies, federal, state, and local."

A.1.2 Relationship with the Populace

Police information flows continuously through the interpersonal information network established during the conduct of security operations in support of full spectrum operations and during LE operations in support of installations. Information can originate from a multitude of sources and may be obtained through deliberate collection efforts or passive collection resulting from police engagement and observation. Whether collected for a specific LE purpose or as a byproduct of daily interaction and situational awareness by NSF, information collected by police personnel provides valuable insight to the local environment. Police information and police intelligence enhance situational understanding, provide critical information that can fill gaps in IR, and add to the overall common operational picture.

The relationship between the police and the population is critical in the ability to interact with and operate around the local population. Just as mistrust and a weak social order can help criminals and terrorists, a strong relationship between the population, police, and security forces is critical to assisting with investigations, understanding the social order, and defeating criminal networks. The relationship between the police and the populace in many areas is often not as strong as it should be to enable full support to LE forces. Examples of police corruption, excessive use of force, preferential treatment, and evidence mishandling have received regional and national exposure. Evidence of these actions, even if they are very isolated incidents, can degrade that relationship. Rebuilding public confidence in police forces requires both organized public awareness efforts and continuous professionalism on the part of security forces. Navy security and LE operations are guided by the principles of policing.

The principles of military policing are:

1. Prevention

2. Public support

3. Restraint

4. Legitimacy

5. Transparency

6. Assessment.

Navy Security Forces are subject to the same forces that build public confidence and erode it. On Navy installations, NSF should always be cognizant of the importance of a positive relationship with Sailors, family members, residents, installation workers, and the general public.

A.1.3 Identifying Collection Assets

The CO, supported by the SO and staff, selects and prepares collection assets based on their capabilities and limitations. NSF are trained collectors and highly adaptable to any collection plan. These personnel operate in direct contact with the local population, allowing them to identify, assess, and interact with potential sources of information. NSF LE personnel can also effectively collect information as a deliberate collection mission or concurrent with the conduct of other missions and functions. Information can be collected actively (through direct observation and/or engagement with specific targeted personnel) or passively (by observing and listening to the surrounding environment and personnel). These collection of activities span all environments and spectrums, from routine and relatively stable environments associated with LE in support of installations to the extreme instability of combat operations.

On Navy installations, NSF and NCIS personnel are the lead for targeting, collecting, and interdicting against a broad range of threat activities, including terrorism, organized crime, contraband trafficking, and other illegal activities. Collection of police information in this environment is conducted by NSF and NCIS assets through deliberate surveillance missions or through missions tasked to LE patrols and conducted in the course of their routine patrol or investigative activities.

A.1.4 Reporting and Recording Collected Information

Established debriefing procedures to gather collected information, including debriefing patrols with no deliberate collection mission to gain information gathered during execution of normal operations, is critical to the collection process. Recording and systematically cataloging information obtained by assigned collection assets and routine patrols are critical to the PIO process and may fill important gaps in the overall common operating picture. Police information may be recorded manually or by direct input into CLEOC. The information is compiled for assessment and analysis by the staff and NSF assigned police intelligence analyst responsibilities. If raw police information or police intelligence derived from rapid analysis of the information is identified, it is fed into the intelligence process (as applicable). Due to the restrictions placed on information gathered on U.S. citizens, police intelligence must be provided to the security department operations section for further action when the installation is in the United States and its territories.

A.1.5 Granting Access and Sharing Rights

Granting access to databases, information, or intelligence ensures that personnel, units, or organizations with requirements and legal authorization for access to police information and intelligence are provided the means to obtain the required information. Police information and intelligence may be stored in established LE-sensitive, classified, and unclassified databases and associated programs, networks, systems, and other Web-based collaborative environments. Every effort will be made to ensure that LE agencies operating in the area have access, as appropriate and within legal and policy guidelines. Access and sharing rights are granted through applicable national agencies and according to applicable regulations, policies, and procedures for personnel accesses and clearances, individual system accreditation, specialized training for access and systems or database use, and special security procedures and enforcement.

Sharing access is primarily the result of establishing a collaborative environment for transferring police information and intelligence. Advances in database technology, combined with an explosion in information sharing and networking among police agencies, has resulted in the development and expansion of these robust information repositories. Navy LE personnel continue to access the National Crime Information Center (NCIC) database, but can also turn to databases containing fugitive information from corrections systems and terrorist threat information from the United States Department of Homeland Security (DHS) and Federal Bureau of Investigation (FBI) systems. DOD proprietary automation systems, such as CLEOC, greatly improve interoperability and eliminate seams that criminal and other threats might otherwise exploit. Access to local, theater, DOD, non-DOD, and commercial databases allow analysts to leverage stored knowledge on topics ranging from basic demographics to threat characteristic information. The challenge for an analyst is to gain an understanding of the structure, contents, strengths, and weaknesses of the database, regardless of the database type.

A.2 INFORMATION REQUIREMENTS

Collection includes the activities required to gather and report police information to answer IR. Collection may involve gathering new relevant data and raw information or exploiting existing police intelligence products. Effective collection efforts are generated and driven by the operations process. They are planned, focused, and directed, based on the CCIR, threat assessments, police intelligence, and investigative requirements. Success comes from integrating information gathered during NSF collection and assessment activities and conducting analysis and fusion with other sources of information to answer IR.

The COs designate the most important, time-sensitive items of intelligence they need to collect and protect as CCIR; PIO anticipates and responds to the CCIR. Police intelligence analysts monitor the PIR and subordinate IR for information of value in determining criminal COAs.

A.2.1 Collection of Police Information

Collection is a continuous activity. The Security Officer, staff, and NCIS personnel identify gaps in existing police information and develop IR. In turn, a police information collection plan is developed and targets are nominated for collection against the IR requirement. This collection may be completed by many means, including:

1. NSF police patrols.

2. Police engagement.

3. Reconnaissance, surveillance, and assessments.

4. Criminal investigations.

5. Interviews and LE investigations.

6. Collected evidence (including biometric data and forensic evidence).

7. Data mining, database queries, and use of reach-back centers.

Reliability of the information collected should always be scrutinized to check its viability and credibility. This is especially critical when dealing with individuals providing information, regardless of the environment. Police should be aware of underlying motivations that may drive persons providing information. While conducting LE in support of installations, persons may be motivated to pass information to NSF due to a sense of duty or justice; the military culture is based on values that encourage a sense of duty, honor, and doing what is right. This is beneficial when policing military communities. Others may come forward because they may be complicit in criminal activity and are cooperating in hopes of receiving leniency. Some may seek to obtain revenge against an individual who has done something (whether real or perceived) to slight, hurt, or anger them. These are but a fraction of the possible factors that may motivate members of a population to come forward to LE personnel with information.

A.2.2 Methods and Sources of Information

PIO activities are integrated within each security department function. During the execution of LE, police patrols are arrayed across the installation to perform a myriad of policing, protection, and other missions. The ability of an SO to disperse assets across the installation allows for a significant number of specialized sensors and collectors to gather information required to fulfill CCIR, IR, and other investigative requirements.

A.2.2.1 Active and Passive Collection

Police information can be gathered as a result of passive or active (deliberate) collection efforts during the course of normal LE patrols and activities. Passive collection is the compiling of data or information while engaged in routine missions or LE activities; during passive collection, the NSF patrol is not on a dedicated reconnaissance, assessment, or collection mission. Passive collection occurs every time NSF engage with or observe the people or environment in which they operate. Examples of passive collection include establishing contact with the local population to support rapport; maintaining efforts to clarify and verify information already obtained through observations or other means; or simply observing activity, lack of activity, or other variations from the normal.

A.2.2.2 Police Engagement

Police engagement is a cornerstone to successful long-term police operations. Successful police organizations interact with and gain support from the majority of the population they serve. This holds true for both civilian and military forces in any OE. Police engagement occurs—formally and informally—any time NSF personnel interact with area residents, host nation police and security forces, media personnel, and any of the numerous other avenues that allow personnel to gain and share information about threat and criminal activity in an AO. Data (information) obtained through police engagements must be collected, analyzed, and distributed in a timely fashion to be of maximum value to commanders.

A.2.2.3 Community Leaders

Off-installation community leaders can be a valuable source of information specific to their areas of influence. They will typically have historic knowledge of persons and activities in their cities, towns, neighborhoods, or local military installations. While all information received should be confirmed and vetted, community leaders can provide NSF with valuable information regarding the criminal history in their area (such as activities, persons, and groups), the arrival of new persons or the emergence of groups, and activities and observations that are "out of the norm." These leaders can also provide insight into the disposition of the local citizens towards law enforcement in the area (such as animosity, levels of trust, and perceptions).

A.2.2.4 Initial Complaints and Contacts

NSF personnel can gain a significant amount of information from initial complaints or calls for response due to specific emergencies or incidents. The initial contact with complainants or individuals at the scene of an incident (such as witnesses, victims, or potential perpetrators) can result in valuable pieces of information that may not be available with the passage of time. These circumstances provide witnesses, victims, and perhaps contact with potential perpetrators that have recent memory of an event or a valuable observation. It is important that this information is captured and documented as quickly, thoroughly, and accurately as possible.

A.2.2.5 Surveillance

NSF may be required to conduct surveillance focused on observing specific criminal or threat targets and gathering required information. LE surveillance may be required to confirm suspected criminal activity, establish association of a suspected criminal in terms of time and place, and confirm association between persons, groups, or entities. Surveillance may be physical observation of a person or location, visual surveillance by remote video equipment, or audio surveillance via a myriad of technologies employed to intercept audio evidence. LE surveillance is typically associated with specific criminal investigations. However, LE surveillance may also be performed to conduct assessments, such as traffic studies, physical security assessments, or other security and protection requirements.

A.2.2.6 Interviews and Law Enforcement Interrogations

Although physical evidence, records, and recordings can often provide critical bits of information about an incident, there is almost always significant benefit in asking questions of persons who have some knowledge of an incident (including preparation and aftermath activity). There are three categories of question-and-answer sessions: interviews, LE interrogations, and tactical interrogations. Interviews are conducted with persons who may or may not have information important to an incident and are, by general definition, not confrontational. Interviews are used by LE personnel during the initial-response phase to determine facts regarding an incident. They are also conducted by NSF personnel in an effort to gain background or corroborative information. LE interrogations are conducted by security department CCI's, NCIS special agents, or host nation security forces with individuals suspected of a crime in which some type of prosecutorial outcome is expected. LE interrogations are generally more confrontational than interviews.

A.2.2.7 Informants and Law Enforcement Sources

In some circumstances, NCIS special agents may attempt to gain recurring access and insight into the workings of a criminal or terrorist network through the use of informants. At other times, they will seek similar access to an organization that may knowingly or unknowingly provide support to criminals or terrorists. Navy NSF frequently obtain information from an LE source who, for a variety of reasons, are willing to provide such information. At times, these persons may be anonymous and available only once or twice. At other times, they will be known to the NSF and may be willing to provide additional information, including information they obtain specifically upon request.

A.2.2.8 Open Source and Public Information

Open source and public information can provide a significant amount of information that may be useful to police intelligence analysts. Trends, patterns, and associations can be determined from open sources, such as newspapers, press releases, and other publications. With the proliferation of data in the public domain over the internet, police intelligence analysts can find significant information for integration and fusion with existing police information and intelligence.

A.2.3 Reporting Collected Information

Information collected is of no use unless it is provided to the appropriate personnel in a timely manner. Following any collection activity, reports must be compiled for the staff, investigator, or commander requiring the information. The location of physical evidence must also be preserved and reported to maintain chain-of-custody requirements and to allow timely reexaminations of other evidence. Appropriate data should be provided to police intelligence analysts supporting LE and investigative operations. Immediate assessment of collected information may lead to the identification that the information has answered an IR or a PIR. When identified, information should be reported to the appropriate staff, commander, investigator, or security officer. Time-sensitive information identified as exceptional should be immediately reported through the appropriate staff and command channels for action. While it is important to produce police intelligence, it is extremely important to share not only police intelligence but also raw information, when appropriate. The value of raw information should not be overlooked; an item of information that is not of particular value to one investigation may be important to an adjacent LE organization at a later date. Terrorists and criminal enterprises have robust information-sharing capabilities. Tactics used successfully in one location may be used elsewhere in a matter of hours or days. Information sharing allows staff and analysts to see a broader spectrum of issues. However, police information or intelligence may be so important that its existence cannot be immediately shared. In some instances, it may be possible to develop a synopsis of information that conceals the method, technique, or source. When sharing such synopsized information, it is also important to provide contact information to allow the receiving element to ask further questions and possibly receive additional information.

A.3 ANALYSIS OF POLICE INFORMATION

The purpose of police intelligence analysis is to answer PIR and other supporting IR and produce police intelligence in support of LE missions and Navy protection operations. Analysis in the context of PIO is conducted from a policing and LE investigative viewpoint and is focused on policing activities, systems, capabilities, and the criminal dimension in the OE. Analysis is one of three continuing activities in the intelligence process (and likewise in PIO) that involve integrating, evaluating, analyzing, and interpreting information from single or multiple sources into a finished product (police intelligence). Police intelligence products must include the needs of the CO, SO or investigator and be timely, accurate, usable, complete, precise, reliable, relevant, predictive, and tailored.

Analysis enables the development and recognition of patterns and relationships. Tools and techniques for analysis provide methods to manage or manipulate those relationships and patterns to draw relevant and accurate conclusions. More simply, analysis is a structured process through which collected information is compared to all other available and relevant information to:

1. Develop theories and form and test hypotheses to prove or disprove accuracy.

2. Differentiate between the actual problem and the symptoms of the problem.

3. Enable the analyst to draw conclusions.

4. Develop adversary or criminal COAs.

Analysis requires manipulating and organizing data into categories that facilitate further study. Patterns, connections, anomalies, and information gaps are assessed during the analysis of police information. The initial hypothesis and data comparison is accomplished by performing the following steps:

1. Observe similarities or regularities.

2. Ask what is significant.

3. Categorize relationships.

4. Ascertain the meaning of relationships or lack of correlation.

5. Identify RFIs and the need for additional SME analyses.

6. Make recommendations for additional collections, to include locations and time constraints.

NSF personnel analyze, synthesize, and develop analytical products based on available information. All analytical techniques use cognitive thought and require analysts to deduce, induce, and infer while working toward conclusions that answer specific IR. Depending on specific requirements and missions, categorizing the effort helps to focus the analyst's efforts toward specific data, considerations, and results. The following paragraphs discuss several analytical focus areas.

A.3.1 Communication Analysis

Communication analysis depicts telephone records, including the analytical review of records reflecting communications (such as telephones, e-mails, pagers, and text messaging) among entities that may be reflective of criminal associations or activity. It may result in identification of the steps required to continue or expand the investigation or study. While communication analyses have long been a key element in the context of LE investigations, advances in technology have elevated the importance, capability, and scope in regard to tracking communications activities of individuals and organizations during investigations. Communications analysis can enable an analyst to document incoming and outgoing calls, locations of phones, date and time of calls, duration of calls, and other communications data. This facilitates identification of communication patterns and associations relative to specific communications equipment.

A.3.2 Crime and Criminal Target Analysis

Crime and criminal target analysis enables the staff and analysts to identify potential criminal targets and crime-conducive conditions, including assessments of vulnerability and relative importance or priority for targeting. A key aspect to performing a crime and criminal target analysis is the determination of the effect desired and the optimal method of targeting. NSF personnel use crime and criminal target analysis to identify criminal targets and crime-conducive conditions and to make recommendations on appropriate engagement methods. Targeting could range from police (information) engagements to the application of nonlethal and lethal force, depending on mission and operational variables.

A.3.3 Crime and Criminal Threat Analysis

Crime and criminal threat analysis is a continuous process of compiling and examining all available information concerning potential criminal threat activities. Criminal and terrorist threat groups or individuals may target U.S. military organizations, elements, installations, or personnel. A criminal threat analysis reviews the factors of a threat group's operational capability, intentions, and activity and the OE in which friendly forces operate. Threat analysis is an essential step in identifying and describing the threat posed by specific group(s) and/or individual(s). Criminal threat analysis techniques are regularly applied to antiterrorism, physical security, and conventional criminal activities. Irregular forces operating in the vicinity of Navy installations against U.S. interests may use criminal and terrorist tactics, techniques, and procedures that also make them viable targets for friendly threat analysis. Crime and criminal analysis supports the production of the crime prevention survey (CPS) and other threat assessments.

A.3.4 Crime Pattern Analysis

Pattern analysis is the process of identifying patterns of activity, association, and events. A basic premise is followed when using this technique: activities, associations, and events occur in identifiable and characteristic patterns. Crime pattern analysis looks at the components of crimes to discern similarities in the areas of time, geography, personnel, victims, and modus operandi. Crime pattern analysis can be critically important when facing a threat in which doctrine or the mode of operation is undeveloped or unknown but is necessary to create a viable threat model. Crime pattern analysis is particularly applicable in LE and investigative applications. Crime pattern analysis can be employed using several different analytical methods. These tools include:

1. Crime and criminal trend analysis.

2. Pattern analysis.

3. Link, association, and network analysis.

4. Flowcharting.

5. Time, event, and theme line charting.

A.3.5 Functional Analysis

Functional analysis is focused on assessing a threat disposition and action for a particular type of operation. Functional analysis is based on the concept that certain operations or tasks are explicitly unique; certain actions or functions must be implicitly performed to accomplish those operations or tasks. The functional analysis provides a framework for understanding how specific threats make use of their capabilities. Functional analysis is applicable regardless of how the threat is characterized. Specific knowledge and training enable analysts to apply the functional analysis process, which effectively addresses specific types of threats. Functional analysis typically consists of the following steps:

1. Determine the threat objective.

2. Determine the functions to be performed to accomplish the identified threat objective.

3. Determine the capabilities available to perform each function.

4. Graphically depict the threat use of each capability.

In the context of the functional analysis, security staff and police intelligence analysts also conduct a criminal threat risk analysis. The purpose of the criminal threat risk analysis is to determine the relative risk that a specific criminal threat poses to military forces, assets, or the population in general. A heightened criminal threat probability typically drives a more rapid and focused action on the part of NSF and other police agencies working in concert.

A.3.6 Financial Crime Analysis

The purpose of a financial crime analysis is to determine the extent to which a person, group, or organization is receiving or benefiting from money obtained from nonlegitimate sources. Financial crime analysis is applicable to many criminal investigations, including organized crime, drug trafficking, human trafficking, and property crime, particularly those involving crimes where money is a motivating factor. This type of analysis is usually performed by NCIS or other federal investigative agencies and focuses on financial and bank records, the development of financial profiles (through net-worth analyses, identifications of sources, and applications of funds), and business records.

A.4 ANALYTICAL TOOLS AND TECHNIQUES TO IDENTIFY TRENDS, PATTERNS, AND ASSOCIATIONS (CRIME PATTERN ANALYSIS)

There are many tools and techniques available to staffs and police intelligence analysts to focus efforts and maximize the effectiveness of analyses. These tools and techniques are used to recognize trends, patterns, and associations. These techniques are not used as singular methods but, rather, are sometimes concurrent and often consecutive activities that complement and enhance each other. Qualitative and quantitative data are used in these techniques. Qualitative data refers to nonnumerical data. This type of data lends itself to content analysis and the identification of historical trends, patterns, and associations. Quantitative data is typically numerical, and analyses of quantitative data are typically statistical in nature. Police intelligence analysts use these tools and techniques to fuse or disparate information in police intelligence products. They are also used to help in developing crime trends and patterns and performing predictive analysis. These tools and techniques help NSF personnel determine what crimes or events are taking place, where they will be located, what time they will occur and, often times, what future activities may occur. These tools may link crimes to threat group activities that may impact the common operational picture, IPB, and CCIR. Throughout the analysis of police information and the production of police intelligence, relevant information and intelligence is provided to the operations and integrating processes.

A.4.1 Establishing Trends (Trend Analysis)

Trend analysis refers to the gathering, sorting, prioritizing, and plotting of historical information. It provides analysts and supported commanding officers a view of how events, elements, and conditions have affected police operations and criminal dimensions in the past. Statistical analysis allows an analyst to extrapolate data to predict future actions or occurrences. This historical perspective provides continuous insights for developing coherent possible and/or probable COAs for the criminal threat and the ability to predict specific occurrences. The results of trend analysis are sometimes referred to as statistical intelligence. In the context of PIO, statistical intelligence refers to data collected from police reports, raw data files, and other historical data assembled into useable maps (or other geospatial products), charts, and graphs. This information is used to indicate past crimes and trends, patterns, or associations. This information must be maintained and updated to be effective.

Police intelligence resulting from trend analysis is the baseline that analysts and units should use as a statistical point of reference for future analyses. Trends can be depicted in many different formats, to include graphs, maps (or other geospatial products), and narrative summaries. Ideally, trend analysis products should be depicted visually and in a report format. A trend analysis is extremely useful for:

1. Specific occurrences.

2. Traffic accidents.

3. Driving while intoxicated and other alcohol-related incidents.

4. Juvenile crimes.

5. Assaults (including simple, aggravated, and domestic incidents).

6. Sex crimes.

7. Suicide.

8. Drug offenses.

9. Homicide.

10. Larcenies.

11. Gang activities.

12. Security-related incidents (perimeter breaches, unauthorized entry, exclusion area violations).

13. Offenses by specific persons (persons with a criminal history).

14. Locations and times of specific offenses.

15. Complaints against the police.

16. Number and type of citations.

17. Calls for assistance.

18. Response times.

19. Special-event attendance statistics.

20. Traffic flow.

21. Specific intersections or roadways.

22. Entry control points and traffic control points.

23. Traffic peaks (including daily, seasonal, holiday and special events).

Comparisons of the recorded historical police and criminal events and associated trends derived through statistical trend analysis can provide clues to criminal and threat capabilities, modes of operation, and activities in relation to time and location. Police intelligence derived from trend analysis enables the redistribution of police assets to address specific policing problems. Trend analysis can also determine organizational problem areas and facilitate organizational adjustments or changes to improve operations.

A.4.2 Identifying Patterns (Pattern Analysis)

Pattern analysis helps an analyst identify indicators of threat activity. A pattern analysis is based on the premise that activities conducted by individuals, groups, or organizations tend to be replicated in identifiable ways. A thorough analysis of seemingly random events can result in the identification of certain characteristic patterns. Pattern recognition defines the ability of an analyst to detect and impose patterns on random events, allowing for the separation of relevant information from irrelevant information. Pattern recognition can enable an analyst to make assumptions and predictions based on previous historical patterns of activity. In the context of PIO, pattern analysis looks for links between crimes and other incidents to reveal similarities and differences that can be used to help predict and prevent future criminal, disruptive, or other threat activities. There are numerous tools and techniques that can be used to display data and establish patterns for analyses. These tools and techniques include:

1. Association and activities matrices. These tools are used to determine associations between persons and activities, organizations, events, addresses, or other variables. Association matrices establish the existence of known or suspected connections between individuals. An association matrix may be reflected as an array of numbers or symbols in which information is stored in columns and rows. Activities matrices do not develop associations between people. (See figure A-1.)

2. Incident maps and overlays. This tool sometimes referred to as a coordinate's register, documents cumulative events that have occurred in the area. This technique focuses on where specific events occur. Incident maps and overlays are a critical tool in geographic distribution analysis. GIS tools can be helpful in producing incident maps with overlay data. (See figure A-2.)

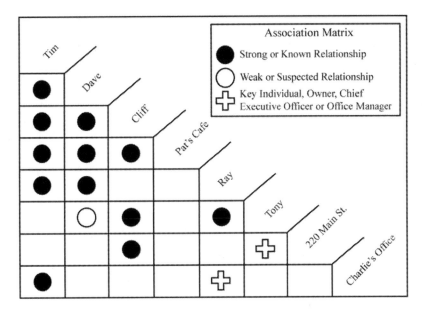

Figure A-1. Example of an Association Matrix

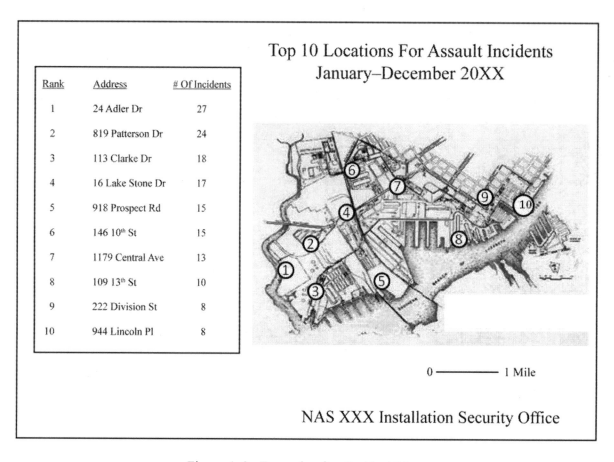

Figure A-2. Example of an Incident Map

3. Pattern analyses plotting. This tool focuses on identifying patterns based on the time and date of occurrences.

A.4.3 Identifying Linkages and Associations (Link and Network Analysis)

Link analysis is a technique used to graphically depict relationships or associations between two or more entities of interest. These relationships or associations may be between persons, contacts, associations, events, activities, locations, organizations, or networks. Link analysis is sometimes referred to as association or network analysis. Police intelligence analysts use link analyses to find and filter data that will locate people; identify ownership of assets; and determine who is involved, how they are involved, and the significance of their association. Link analysis can be especially valuable to active, complex investigations. (See figures A-3 and A-4.)

It provides avenues for further investigation by highlighting associations with known or unknown suspects. Link analysis is normally tailored to a specific investigation; therefore, dissemination is generally restricted to other LE or military personnel acting as part of the same investigation. The main reason for using link analysis is to provide a visual depiction of the activities and relationships relevant to the investigation or operation being conducted. The visual depiction of the network gives meaning to data absent from a visual depiction because it would be too confusing to comprehend. Link analysis is a good analytical tool for generating inferences based on what is known about the current relationships of the known individuals being targeted. The network charting tool can depict the ever-changing alliances and relationships relevant to the investigation or operation.

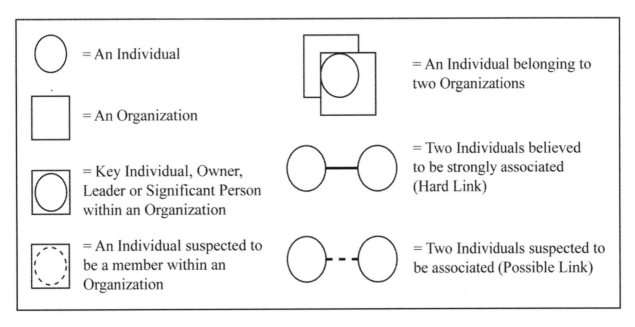

Figure A-3. Example of Standard Link Analysis Symbology

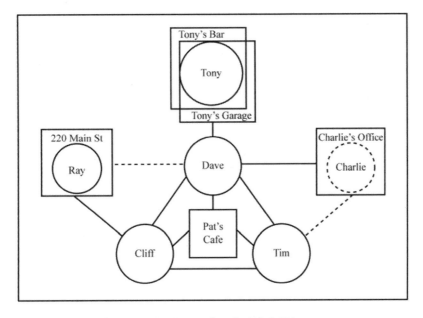

Figure A-4. Example of a Link Diagram

A.4.4 Flowcharting

Flowcharting is a series of analytical techniques that describes and isolates the distribution pattern of a criminal organization, their mode of operation, and the chronology of crime-related activities. Flowcharting allows an analyst to isolate associations and patterns identified through previous analysis techniques to depict a specific person, organization, entity association, or activity without the extraneous information that may have been present in earlier analysis techniques. The flowchart may also show gaps in time that need to be accounted for. When combined with ventures and link analysis charts, a flow chart can assist personnel in understanding relationships and where all of the involved associates fit in to the scheme of the criminal or terrorist enterprise. Some flowcharting techniques include:

1. Activity. This technique depicts the key activities and modes of operation of an individual, organization, or group. Activity flow analysis is used to view criminal actions and identify modus operandi to determine

likely suspects. Most criminals will leave unique indicators when committing a crime. These indicators are specific details, common to the specific criminal or organization, and may include details regarding types of weapons, notes, vehicles, targets, or number of people involved.

2. Time event and theme line charts. These tools establish chronological records of activities or related events. The charts may reflect activities of individuals or groups and depict large-scale patterns of activity. Figure A-5 shows an example standard symbology used in time event charts, while figure A-6 shows a basic example of a time event chart.

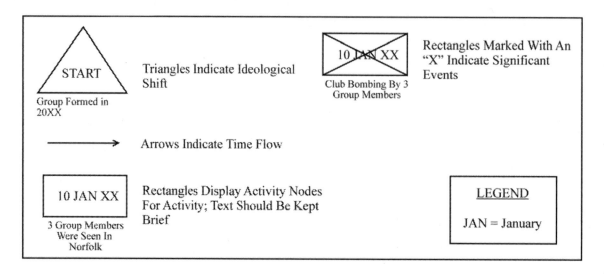

Figure A-5. Example of Standard Time Event Symbology

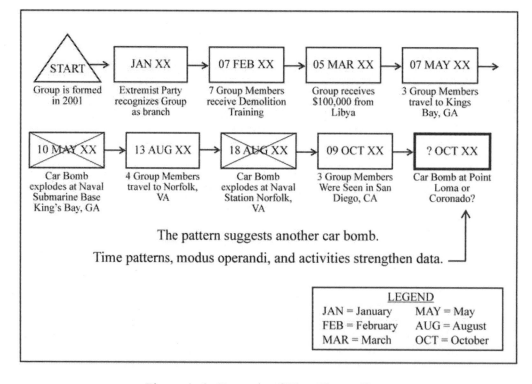

Figure A-6. Example of Time Event Chart

A.4.5 Compiling Statistical Data

Statistical data can be drawn from diverse sources and depicted in numerous manners. Statistical data can be very useful in determining frequencies, trends, and distributions; however, the data also has limitations. Analysts and users must be aware of the limitations and resist the desire to infer more than the presented data can accurately portray. Statistical data may be presented in charts displaying:

1. Frequencies—through the use of bar or pie charts.

2. Trends—through the use of bar and line charts.

3. Distributions—through the use of GIS, trend lines, or other formats.

A.4.6 Crime and Criminal Threat Analysis

Compiling, examining, and reexamining all available information concerning potential threat activities is a continuous process. Threat analysis is conducted by intelligence and LE organizations to determine and monitor current and potential threats. It is an integral element in the production of a threat assessment. Threat assessments must be current to be relevant. Intelligence and criminal information, threat information, and asset vulnerabilities are all considered when conducting a threat analysis. Intelligence and criminal information provide data on the goals, methods of operation, techniques, strategies, tactics, and targets of individuals and groups. Threat information can lead to the identification of criminals and criminal groups. A threat analysis must be a continuous activity to account for inevitable changes in the OE. As vulnerabilities are reduced in some areas, they may increase in others. Threat elements assess their targets in relation to one another. An increase in the security posture of one asset may increase the attractiveness of another asset as a target, even though the asset has not reduced its security. Changes in mission, tasks, and personnel also may have an impact on the status of the current threat analysis. Criminal and terrorist threat groups or individuals may target Navy installations, organizations, elements, or personnel. A threat analysis reviews the factors of a threat group's operational capabilities, intentions, and activities and the OE in which friendly and threat forces operate. Threat analysis is an essential step in identifying and describing the threat posed by specific groups and/or individuals. Threat analysis is most typically associated with terrorist activity, but the same techniques are applied to conventional criminal activities. A threat assessment integrates a threat analysis with criticality and vulnerability assessments that are required for prioritization of assets by commanding officers and installation security officers to counter threat activities and associated risks. Vulnerability and criticality information helps the analyst to identify security weaknesses and potential high-risk targets.

The ultimate goal of a crime and criminal target analysis as it relates to PIO is to identify criminal and threat persons, groups, or organizations; identify historic and current crime trends and predict future trends and activities that enable targeting decisions; develop investigative leads through the identification of trends, patterns, and associations; and make target recommendations. Crime and criminal target analyses and subsequent support to operations occur in all environments.

INTENTIONALLY BLANK

APPENDIX B

Use of Force

B.1 USE-OF-FORCE POLICY

All NSF personnel engaged in LE, security, and force protection duties shall only use the amount of force reasonably necessary to carry out their duties. Frequently, NSF personnel may need to use force to perform their assigned duties. As such, the appropriate use of force shall be applied without prejudice within the law, Navy policies, and guidance. To ensure compliance, NSF shall receive the appropriate training in its application. This NTTP is designed to assist in the specific use of force, including deadly force training and techniques, used by all NSF personnel.

In cases warranting the use of force, NSF personnel may only use that force reasonably necessary to control or stop unlawful resistance to reach the objective and prevent the unlawful commission of a serious offense. Always use the following principles.

1. Force must be reasonable in intensity, duration, and magnitude.

2. There is no requirement to delay force or sequentially increase force to resolve a situation or threat. NSF personnel will attempt to de-escalate applied force if the situation and circumstances permit. NSF personnel will warn persons and give the opportunity to withdraw or cease threatening actions when the situation or circumstances permit.

3. Warning shots are prohibited. Warning shots are authorized from U.S. Navy and naval Service vessels and piers in accordance with Chairman of the Joint Chiefs of Staff Instruction 3121.01B, Standing Rules of Engagement/Standing Rules for the Use of Force for U.S. Forces, 13 June 2005.

4. Firearms shall not be fired solely to disable a moving vehicle. When deemed as a threat to DOD assets or persons' lives, NSF personnel shall use reasonably necessary force and caution when firearms are directed at a vehicle borne threat.

5. If excessive force is used in discharging assigned responsibilities, NSF shall be subject to administrative or judicial action IAW Article 92 of the UCMJ or U.S., local, or HN laws.

All Navy use-of-force policies and guidelines are applicable in overseas areas providing they satisfy all applicable provisions of international agreements, SOFAs, MOAs, or arrangements relating to LE and security matters and established ROE.

B.2 USE-OF-FORCE PRINCIPLES

NSF engaged in the protection of personnel and resources have the authority to employ force on subjects, including the use of NLW and firearms if required. Before using force, NSF are required to possess the knowledge and skills necessary to assess acts or threats and respond in a reasonable manner. These objective principles enable NSF to make reasonable and judicious judgments that support the overriding thought of their safety and the safety and security of personnel and resources they are protecting. The decision to use force can have severe consequences that may result in the loss of life or injury of another person. NSF must remember, once force is employed, that force is irreversible. Alternatively, hesitancy or inaction when force is necessary may be devastating to the safety and security of personnel and resources being protected. Therefore, this section discusses mental preparedness, basic principles of the use of force, the use-of-force model, and the tools and tactics available within the Navy.

B.3 CONTINUUM OF FORCE

NSF shall only use the amount of force reasonably necessary to carry out their duties. The following use-of-force continuum is designed to provide an overview and visual representation of the force options available to NSF. It is a fluid instrument that attempts to embody the dynamics of a confrontation.

1. The standard for evaluating NSF use of force shall be its reasonableness under the facts and circumstances known to the NSF member at the time.

2. A number of factors should be taken into consideration when NSF select force options and when commands evaluate whether NSF have used reasonable force. Examples of factors that may affect NSF force option selection include but are not limited to:

   a. Subject factors (age, size, relative strength, skill level, injury/exhaustion, number of patrol officers versus number of subjects).

   b. Influence of drugs or alcohol.

   c. Proximity to weapons.

   d. Availability of other force options.

   e. Seriousness of the offense in question.

   f. Other urgent circumstances.

3. NSF need not attempt to gain control over an individual by use of the lowest level of force on the continuum when reason dictates and the officer can articulate that a higher level of force is reasonable. Likewise, the skipping of steps may be appropriate given the resistance or circumstance encountered.

4. The following category descriptions are nonexclusive and are intended to serve as illustrations of actions that fall within the various levels.

5. Actions of subject (as reasonably perceived by the officer or based on the officer's reasonable perception) include:

   a. Cooperative: Subject is cooperative and complies with verbal commands or other directions. The likelihood of a physical response by the subject is low. The subject can be controlled through the presence of NSF and verbal skills.

   b. Nonresponsive or uncooperative: Subject fails to respond to verbal commands or other directions.

   c. Passive or low-level resistance: Subject is passively or defensively resisting a patrol officer's authority and direction. Includes verbal or physical cues of noncompliance. Requires some degree of physical contact to elicit compliance.

   d. Active resistance or aggression: Subject exhibits physical defiance and is attempting to interfere with NSF actions by inflicting pain or physical injury on the patrol officer without the use of a weapon or object.

   e. Assault or threat of assault: Subject assumes a fighting stance, charges an NSF, or verbally or physically indicates intent to commit an assault.

f. Life-threatening assault or assault likely to cause great bodily harm: Subject commits an attack using an object, a weapon, or an empty-hand assault wherein the NSF reasonably believes that the assault will result in serious physical injury and/or death.

6. Response options include:

a. Professional presence, verbalization, restraining, and detaining:

(1) Includes display of authority as a NSF and such nonverbal means of communication as body language, demeanor, and manner of approaching.

(2) Verbalization involves the direction and commands given to the subject.

(3) Restraining and detaining include an officer laying hands on a subject with the intention of gaining control of the subject. Also included in this level would be the application of temporary restraining devices such as handcuffs and leg restraints.

b. Compliance techniques: Includes joint manipulations, pressure-point applications, take-down techniques, and the use of intermediate weapons in control-type configurations.

c. Intermediate force: Includes chemical agents such as oleoresin capsicum (OC)–based products; the use of impact weapons in an impact mode; patrol MWDs; and the use of personal weapons such as hands, feet, elbows, and knees to strike a subject.

d. Deadly Force: Force that a person uses causing, or that a person knows or should know would create a substantial risk of causing death or serious bodily injury.

B.4 NONLETHAL WEAPONS

NSF personnel may be armed with weapons or equipment other than firearms that when applied, even though their intended purpose is nonlethal, could cause death or serious bodily harm. DODD 3000.3, Policy for Non-lethal Weapons, maintains:

1. The possession of NLW shall not limit NSF personnel so equipped from using deadly force. Neither the presence nor the potential effect of NLW constitutes an obligation for their employment or a higher standard for employment of force than provided in SECNAVINST 5500.29 (series), Use of Deadly Force and the Carrying of Firearms by Personnel of the Department of the Navy in Conjunction with Law Enforcement, Security Duties, and Personnel Protection.

2. The use of NLW shall not be required to have a zero probability of producing fatalities or permanent injuries. However, while complete avoidance of these effects is not guaranteed or expected, when properly employed NLW should significantly reduce them. The following provisions apply:

a. Handcuffs. Handcuffs shall be securely fashioned but not so tightly as to cause the individual injury or pain. When in use, handcuffs shall be double-locked and checked periodically to ensure they are not causing injuries. Subjects shall not be handcuffed to vehicles. Handcuffing to other objects, such as interrogation desks, should be considered only when subjects are deemed to be a danger to themselves or the interviewer. These precautions are also applicable to the use of flex cuffs and leg irons. The removal of handcuffs can present as many possible safety threats as applying them. It is important to follow a removal procedure that keeps the subject off balance and discourages an escape attempt or assault. It is strongly recommended that other NSF be present when handcuffs are removed.

b. Batons. NSF personnel must avoid intentionally striking combative subjects in the head or other bony body parts with batons, except when circumstances justify the use of deadly force. Expandable batons may be issued only after appropriate training has been provided.

B.5 RIOT CONTROL AGENTS AND OLEORESIN CAPSICUM SPRAY

NSF may use riot-control agents (RCAs) to subdue a subject in self-defense or for protection of a third party if circumstances warrant their use. Appendix K.5 contains additional guidance concerning the use of RCAs.

1. OC spray is considered an RCA and may be used only when the use of such force is necessary and reasonable under the circumstance at the time. OC may be used only in the degree/amount reasonably necessary to achieve a lawful purpose (limited to minimal amount necessary to establish control of the offender).

2. OC spray shall not be used for the purpose of aiding an interrogation, punishment, or causing unnecessary pain or discomfort.

3. OC spray is considered an RCA, and its use in war is subject to limitations imposed by the Chemical Weapons Convention. U.S. armed forces are prohibited from using any RCA in war unless the President (POTUS) approves such use in advance. All use of force, including the use of RCAs, whether in war or operations other than war, must be in consonance with the ROE or use-of-force policy applicable to that operation. The use of RCAs outside a war zone is authorized as prescribed for peacetime.

B.6 USE-OF-FORCE NOTIFICATION AND PROCEDURES

NSF shall immediately notify their supervisor whenever a level of force above restrain/detain is used or concerning any incident in which an injury or complaint of injury occurs during the course of contact with a subject, unless urgent circumstances delay the notification.

1. NSF shall obtain medical assistance for subjects who have sustained injuries or complained of injury or who have been rendered unconscious.

2. NSF shall document the use of force in detail in the narrative section of the IR.

3. Upon being notified of an incident involving NSF use of force, supervisors shall:

   a. Make voice reports to the SO and CDO on the level of force used.

   b. Assess the incident, conduct an investigation, collect evidence, and ascertain witness information.

   c. Promptly prepare a memorandum outlining the circumstances of the use of force and send it via chain of command to the SO.

   d. Ensure that all reports have been completed and submitted.

B.7 ARMING NAVY SECURITY FORCE PERSONNEL

The authority to arm NSF personnel is vested in the CO by U.S. Navy Regulations, 10 U.S. Code (U.S.C.) 1585, and SECNAVINST 5500.29 (series), Use of Deadly Force and the Carrying of Firearms by Personnel of the Department of the Navy in Connection with Law Enforcement, Security Duties, and Personal Protection, or, in overseas locations, as governed by SOFAs. Once the determination is made to arm, weapons shall be carried loaded at all times (Condition 1) IAW NTRP 3-07.2.2, Force Protection Weapons Handling Standard Procedures and Guidelines.

No contract guard shall bear firearms onboard a Navy installation until written certification of qualification meeting Navy standards is provided by the contractor and the guard has successfully completed training in the use of force and ROE. In addition, contractors must comply with provisions prescribed by the state/country in which the contract is administered, including current licensing and permit requirements as needed for the individual or company based on the use of deadly force criminal/civil liabilities.

Qualified personnel shall be issued an Authorization to Carry Firearms Form (OPNAV 5512/2), which must be in their possession while carrying a firearm.

B.8 WEAPONS PROCEDURES

1. NSF shall not use, carry, or have in their possession personal weapons while in the performance of assigned duties. NCIS special agents are authorized to carry nongovernment-issued handguns for use in the performance of duties as granted in SECNAVINST 5500.29 (series), Use of Deadly Force and the Carrying of Firearms by Personnel of the Department of the Navy in Conjunction with Law Enforcement, Security Duties, and Personal Protection.

2. NSF shall not carry government-owned weapons when off duty, nor shall they keep government-owned weapons in private residences, either on or off the installation. Government-owned weapons shall be stored only in approved security containers or armories per OPNAVINST 5530.13 (series), Department of the Navy Physical Security Instruction for Conventional Arms, Ammunition, and Explosives (AA&E). Weapons must be returned to approved storage after completion of duty or training.

3. NSF may carry government-owned weapons off base when in a duty status. Security supervisors must ensure compliance with applicable federal and local statutes and SOFAs.

4. NSF shall use only ammunition obtained through Navy supply sources in government-owned weapons.

5. NSF shall not carry unloaded weapons while on duty except for safety reasons while on the firing range or while participating in training exercises.

6. When NSF weapons are stored, weapons storage facilities shall meet the requirements of OPNAVINST 5530.13 (series), Department of the Navy Physical Security Instruction for Conventional Arms, Ammunition, and Explosives (AA&E).

7. NSF vehicle patrols shall secure shotguns in an approved shotgun-locking mount or locked in the patrol vehicle's trunk in such a manner as to prevent shifting/movement of the weapon.

8. NSF should not draw his or her sidearm unless (additional NSF guidance regarding use-of-force guidance can be found in attachment 1 to Appendix S):

   a. There is a reason to believe that it might be needed to protect his or her life or that of another.

   b. Urgent circumstances exist that create a reasonable expectation that deadly force may become necessary under the circumstances.

9. Any time NSF personnel draw their weapon, they should fully document the circumstances IAW the security department's RRP.

10. Loss or theft of firearms or ammunition shall be reported as follows:

    a. NSF personnel shall immediately report the loss or theft of a duty weapon or ammunition to the patrol/post supervisor or watch commander, who shall then report it to the SO.

    b. Discovery of a loss or theft of a NSF weapon from a ready-for-issue (RFI) storage area shall be immediately reported to the senior NSF supervisor on duty, up to the SO.

    c. Upon receiving the report of a lost or stolen weapon, the senior NSF supervisor on duty shall ensure that appropriate RRP are followed regarding initial investigation and notification. Reports shall be made IAW OPNAVINST 3100.6 (series), Special Incident Reporting (OPREP-3) Procedures.

11. NSF shall report any improper discharge of a firearm or any instance of mishandling of firearms by NSF personnel to their supervisor or watch commander, who shall ensure that appropriate RRP are followed regarding initial investigation and notification. All such instances of discharge or mishandling shall be immediately reported to the SO and the Naval Safety Center.

12. Government credentials and badges do not entitle NSF members to carry a weapon in an off-duty status.

B.9 DETENTION CELLS

The operation of detention cells is the responsibility of the SO. When a detention cell is available, the following standards shall apply:

1. Detention cells must be certified IAW OPNAVINST 1640.9 (series), Guide for the Operation and Administration of Detention Facilities, which specifies the requirements for processing of detainees and prisoners.

2. SOPs will be prepared.

3. Only designated personnel shall be allowed into the detention cell area. Personnel working in the area shall not be armed with any weapon, including firearms, RCAs, batons, or other such devices, unless in the performance of duties, such as during prisoner disturbance.

B.10 RULES OF ENGAGEMENT

1. ROE apply to U.S. forces during military attacks against the United States and during all military operations, contingencies, and terrorist attacks occurring outside the territorial jurisdiction of the United States. The territorial jurisdiction of the United States includes the 50 states, the commonwealths of Puerto Rico and Northern Marianas, and U.S. possessions and territories.

2. POTUS and the Secretary of Defense (SecDef) approve ROE for U.S. forces. The Joint Staff, Joint Operations Division (J-3), is responsible for the maintenance of these ROE through CJCSI 3121.01 (series), Standing Rules of Engagement and Standing Rules for the Use of Force for U.S. Forces (U), and supplementary combatant commander guidance.

3. Commanders at every echelon are responsible for establishing ROE for mission accomplishment that comply with ROE of senior commanders and standing rules of engagement (SROE). The SROE differentiate between the use of force for self-defense and for mission accomplishment. Commanders have the inherent authority and obligation to use all necessary means available and to take all appropriate actions in the self-defense of their unit and other U.S. forces in the vicinity.

APPENDIX C

Patrol Concepts

C.1 POLICY

COs are responsible for the good order and discipline of their installations through LE and security patrols. The primary duties of NSF patrols are vast and include, but are not limited to, protecting personnel and property; preventing pilferage; supervising road traffic; enforcing traffic laws and regulations; maintaining good order and discipline of personnel; providing community policing and assistance; and performing security checks and fund escorts. Although these duties occur day-to-day and may appear routine, they must not be accomplished in a haphazard manner. Personnel performing patrol duties must constantly exercise situational awareness. The NSF are the eyes and ears of the CO and must report all suspicious activity.

The objective of patrol operations is to provide 24-hour protection to the citizens; prevent the occurrence of street crimes; respond rapidly to all requests for emergency LE service; improve the criminal apprehension rate by conducting thorough preliminary on-the-scene investigations; reduce traffic congestion and accident hazards through systematic enforcement of traffic laws and ordinances; respond to and investigate motor vehicle accidents; aid victims of accidents; assist citizens in dealing with legal, medical, or social problems through direct crisis intervention and/or by making correct referrals to agencies equipped to deal with such problems; and improve LE/community relations by increasing the quality and quantity of contacts between citizens and LE.

This NTTP covers patrol safety, security and LE patrols, community protection and assistance, associated response procedures, patrol methods and means, patrol distribution, transiting off-installation areas, and reporting times.

All Navy installations have a different number of patrols assigned depending on the size of the installation, required operational capability (ROC) level, and calls for service; however, each installation shall have at least one 24-hour patrol. All inhabited areas of an installation shall be provided with a 15-minute patrol response capability. The patrols shall be linked with each other and the dispatch center by an intrabase radio communications network. If more than one patrol exists on a base, the AO shall be clearly sectored in zones and defined in localized SOP and watch-standing procedures adhered to so as to ensure adequate coverage of the base without redundancy.

C.2 PATROL SAFETY

Navy NSF members have the greatest amount of contact with the public in an uncontrolled environment while on patrol. Navy installations have less violence per capita than society at large. This circumstance may create careless duty behaviors that demonstrate complacent attitudes. Unfortunately, this perspective is wrong. Fatal incidents involving NSF validate the fact that Navy installations are mere microcosms of our society. NSF LE duties place NSF members on the front line against violence. Dispatchers should evaluate responses and dispatch backup patrols as needed. Safety is as paramount for NSF as it is for their civilian counterparts. Often, there is little or no warning before an emergency, crisis, or attack occurs. Always remain alert for the unexpected. NSF, while operating any vehicle shall wear seat belts.

All NSF personnel who operate emergency vehicles will receive the DOT EVOC from a certified instructor at least once every three years. The initial course is 40 hours, and recertification is two or three days. The EVOC and master EVOC curricula are available from the Naval Safety Center, Code 42, 375 A Street, Norfolk, VA 23511-4399. For additional information, visit the Naval Safety Center website at: http://www.safetycenter.navy.mil/.

Authorized training courses to meet NSF skill-level requirements will be posted on NKO. Attendance at nonapproved venues is cost-prohibitive and not authorized. The Center for Security Forces (CENSECFOR) will vet courses recommended to meet NSF skill-set requirements. Use of the Federal Law Enforcement Training Center is not authorized without prior written coordination and approval from CENSECFOR and NCIS Headquarters. State and local training required by law is exempt from prior approval by the CENSECFOR or NCIS.

C.3 TYPES OF PATROLS

The uses of specific types of security and LE patrols are tailored to the needs of the installation. Some situations may call for the use of boat, bicycle, or walking patrols, while others may require motorized patrols. NSF members may perform any or all of the following types of patrols:

1. Security patrols: Single-person mobile patrol units assigned to a specific asset protection zone.

2. Law enforcement: Single patrols responsible for police duties within a patrol zone.

3. Foot and bicycle: Assign these patrols to smaller areas on the installation or those areas with a higher concentration of resources, population, and/or criminal activity. Portable radios link these patrols to the control center and other patrols. MWD teams may supplement foot patrols.

4. Motorized: This method provides mobility and improves the capability to cover a large area of the installation while carrying equipment and personnel. Effective motorized patrols vary their routes. Do not set a pattern or establish a routine.

5. Waterborne: The primary mission of the waterborne patrol shall be to deter unauthorized entry into waterside restricted areas, maintain perimeter surveillance, and intercept intruders prior to their approaching Navy ships in port. Waterborne patrols shall consist of a minimum of two personnel per craft.

6. Reserve: Provide a reserve response capability for emergencies. This may include supervisory patrols, security backup force, or staff personnel.

7. Special purpose: These patrols usually combine motorized, foot, and bike patrols. Special-purpose patrols support unique events (e.g., distinguished visitor conferences, base open house or air shows, unique mission aircraft landings, etc.). Other types of patrols that meet special purposes include horse and all-terrain vehicle patrols.

8. Waterfront Patrols: Security patrols assigned to protect critical assets on the land side of the waterfront, usually on or around piers.

C.4 PRIORITIZING CALLS FOR SERVICE

With the operations tempo in today's Navy, it is not possible for NSF patrols to respond to every call for service. However, installations must organize available resources to give the most efficient service possible. The prioritization of calls has many variables and places a huge responsibility on the dispatch controller.

Senior patrol members may be required to make a quick decision between continuing on an assigned call and responding to a citizen's complaint or other observed event. The senior patrol member's determination should be based on the risk to life and property. When it is impossible for NSF members to respond to a citizen's complaint or an observed event, they should, if circumstances permit, either give direction for obtaining such assistance or start the necessary notifications. Any call for service which NSF are unavailable to respond will be documented in the desk journal.

The following is a suggested but not all-inclusive list of priorities for guidance in responding to calls:

1. Life-threatening emergencies

2. Violent crimes in progress

3. Mission-essential vulnerable assets

4. Other felonies in progress

5. Noncriminal calls with injuries or property damage

6. Other misdemeanors in progress

7. Other felonies not in progress

8. Other misdemeanors not in progress

9. Miscellaneous service calls not involving injury or property damage.

C.5 ASSISTANCE TO MOTORISTS

C.5.1 Motorist Services

Because of the overall danger to the stranded motorist, NSF is expected to offer reasonable help at all times to a motorist who appears to be in need of aid (when in doubt, stop and offer). This expectation applies at all hours of the day, but particularly during the hours of darkness when hazards are greatest. NSF members may be asked to:

1. Get emergency fuel assistance.

2. Get roadside service for a breakdown.

3. Give information and directions.

4. Give first aid and/or medical assistance.

5. Report hazardous conditions.

6. Provide an escort. (Emergency vehicles, particularly ambulances, should not be escorted, except when the driver of the emergency vehicle is not familiar with the destination.)

C.5.2 Disabled-Vehicle Assistance

To prevent the appearance of preferential treatment to commercial establishments, motorists should choose the wrecker or roadside service or ambulance service. If the question arises, caution the motorist that the Navy is not obligated to give towing services or pay for towing and storage costs. Do not use the patrol vehicle to push or pull any vehicle for the purpose of getting it started. Additionally, the patrol vehicle shall not be used to jump-start any vehicle. Conversely, privately owned vehicles shall not be used to jump start Navy vehicles. NSF shall not provide vehicle unlock services. Members may (using discretion) transport stranded motorists to the nearest on-base location where assistance can be obtained. Call in the departure and arrival times and starting and ending odometer readings to the dispatch center.

C.6 APPROACHING SUSPICIOUS PERSONS

Navy NSF patrols are unable to predict which persons may react violently when confronted with LE. Therefore, patrols that are negligent in maintaining awareness for this potential reaction may have tragic results. Each situation warrants vigilance and must be dealt with differently. To ensure that NSF personnel approach suspicious persons in a defensive or cautious posture, the following procedures should be followed. Patrols shall notify the dispatch center of the situation and follow these steps:

1. Call in the number of subjects before approaching them and identify their race, gender, approximate age, and location, as well as a description of their clothing.

2. If possible, have each subject approach separately; do not investigate in a crowd.

3. At night, attempt to direct the patrol vehicle's lights on the subjects; additionally, shine the spotlight on the others.

4. At the NSF member's discretion, each subject/suspect may be checked for wants/warrants through the National Crime Information Center (NCIC)/National Law Enforcement Telecommunications System (NLETS).

5. If warranted, request an additional patrol to assist before approaching.

6. Do not investigate until the assisting patrol arrives.

7. Each patrol should perform a certain part of the investigation:

   a. The first patrol will guard the subjects.

   b. The second officer should perform a pat-down IAW procedures in Appendix H of this NTTP. If the subjects are considered dangerous, patrols should use the prone search position to search.

8. During the interview, patrols are to treat subjects in a courteous manner and inform them of the reasons or probable cause why they were stopped and questioned. Officers are never to question any subject while remaining seated in the LE vehicle with the subject outside.

9. If patrols determine the subject is to be released, they are to obtain the information necessary to complete the OPNAV 5580/21, DON Field Interview (FI) Card.

C.7 USE OF EMERGENCY EQUIPMENT

Use emergency equipment only when directed by a supervisor, by the nature of the dispatched assignment, or when appropriate. The use of emergency equipment is not justification for unsafe driving. Additionally, NSF patrols should routinely test the equipment, using locally produced RRP, to verify that it is in proper working order.

1. Emergency (red/blue) lights.

   a. This equipment is used to signal other users of the traffic-way that emergency conditions exist and the right-of-way should be relinquished to the patrol vehicle. The light is used to signal violators to drive to the extreme right of the roadway and stop. If both the NSF and violator's vehicles are parked off the roadway, the patrol officer and/or violator can stand or walk between the traffic side of their vehicles and the roadway.

   b. NSF members cannot safely assume the light will be sufficient to ensure the right-of-way. Laws exist giving the right-of-way to emergency vehicles when emergency conditions exist. Emergency lights may be used in the following circumstances:

      (1) When stopping traffic violators.

      (2) When assisting motorists parked/stopped in hazardous locations.

      (3) When the patrol vehicle is parked/stopped on the roadway.

    c. Navy NSF members are responsible for any injuries or damage sustained as a result of NSF members' driving behavior that reflects a disregard for the safety of others.

2. Siren.

    a. The siren is frequently used simultaneously with the emergency lights. Use extreme caution when using the siren. The siren may have a startling effect on other drivers on the road, resulting in erratic and unpredictable driving behavior.

    b. The siren should also be used to signal violators to drive to the right of the road when other means of attracting the violator's attention have failed.

    c. NSF members should use the siren based on existing traffic and roadway conditions and the urgency of early arrival. For instance, in traveling to the scene of an emergency, members should use the siren at intersections to alert traffic, but it is sometimes not essential in areas where access to the traffic-way is limited and other traffic is minimal.

3. Emergency lights and siren. Emergency lights and siren in combination should be used in the following circumstances:

    a. Pursuit situations.

    b. When responding to an emergency.

    c. If necessary, to violate traffic regulations when responding to a crime in progress.

4. Spotlights.

    a. Spotlights should be used to aid patrols when hazardous conditions exist in dealing with known or suspected felons. For example, following a traffic stop of a known felon, the spotlight should be used to illuminate the interior of the violator's car so all occupants are kept within view and at a distinct disadvantage when looking back toward the patrol vehicle and patrol officer. In this situation, NSF members should exercise care in remaining behind the spotlight so they are not at the same disadvantage by being silhouetted by the light.

    b. Spotlights should not be used to signal violators to stop, due to the possibility of temporarily blinding the violator and other drivers with the glare created by the spotlight.

5. Public address systems.

    a. Public address (PA) systems are particularly valuable when stopping a traffic violator. The desired actions of the violator can be directed from a safe distance, minimizing hazards to the NSF member.

    b. The PA system is also invaluable in directing persons when unusual conditions exist, such as when a street is temporarily obstructed, alerting pedestrians to hazardous conditions or elements, and communicating with other persons concerned with relieving the emergency conditions.

6. Response procedures. Patrols are usually the first authoritative official to arrive on the scene of an incident. There are many incidents to which patrols may be directed to respond. The patrol officer must be thoroughly knowledgeable in correct response procedures. In response to any incident, the patrol must ensure its safe arrival to perform the duties assigned. The patrol shall then preserve the scene and maintain communications with the NSF supervisor and other patrols, giving information on the status of the incident.

    a. Upon initial response to an incident, ensure the patrol vehicle is properly parked and locked. Avoid parking the vehicle where it may block emergency services such as fire department or an ambulance responding to the incident.

b. Upon arrival, ensure the safety of the scene to preclude further injury or accident. After arrival, the patrol shall first attend to any injured and then preserve the scene for evidence.

c. The dispatch center and other patrols should be kept current on the status of the situation.

d. Witnesses at the scene shall be identified and asked to remain to provide statements.

C.8 RULES FOR PURSUIT DRIVING

Pursuit driving may not always mean driving at high speeds. When engaged in pursuit driving, the patrol officer must remember that the sooner the subject is stopped or apprehended, the less chance there is for an accident to happen. However, patrols must be aware not to be reckless and endanger the public as a result of driving techniques used in pursuit.

1. Pursuits at high speeds are justified only when NSF members know or have a reasonable belief that the violator has committed or attempted to commit a life-threatening offense or felony. This includes offenses that involve an action or threatened attack that NSF members have reasonable cause to believe has resulted in (or is likely to result in) death or serious bodily injury (e.g., murder, attempted murder, kidnapping, aggravated assault, armed robbery, or arson of an occupied building or resource).

2. If necessary and within local constraints, NSF patrols are permitted to use pursuit driving at moderate speeds to apprehend motor vehicle operators who have committed traffic violations, minor offenses, or felonies not previously addressed. Local RRP should specifically describe any constraints regarding pursuit driving.

3. At no time shall NSF use pursuit driving at speeds that will endanger the public or contribute to the possible loss of control of the vehicle.

4. The responsibility for making the decision to pursue an offender and the method used lies with the individual NSF patrol. However, if a pursuit is initiated, the SO, WC, or patrol supervisor shall monitor and may terminate the pursuit at any time if he or she feels it is in the best interest of safety. If pursuit driving is determined to be necessary, the following factors must be considered:

   a. The degree of danger to the public.

   b. Experience and training of the pursuit vehicle's operator.

   c. Weather, visibility, and road conditions.

   d. Pursuit vehicle characteristics.

   e. Present and potential roadway obstacles.

   f. Facilities located along the pursuit route (e.g., schools, hospital, shopping centers, etc.).

   g. Dangerous or potentially dangerous intersections along the pursuit route.

   h. If pursuing a felony suspect off the installation, detailed descriptions of the vehicle, occupants, and suspected offense shall be provided immediately to local law enforcement for their assistance. NSF shall cease off-installation pursuit once local law enforcement assumes pursuit responsibilities. The CO should establish local RRP for NSF and an agreement with local authorities that define the circumstances that warrant an off-installation pursuit, create communication channels to effect immediate reporting of a pursuit to local authorities, and establish procedures to minimize risk to the local populace prior to authorizing pursuits off-installation. Pursuits off an installation must adhere to state and local laws. The RRP should address the use of force and other measures, such as implementing

ECP closures, the use of pop-up bollards, and roadblocks. The CO shall also determine policies regarding pursuit by civilian LE onto the installation after consultation with the SJA or appropriate legal counsel and adjacent jurisdictions.

5. Pursuit driving actions.

   a. Using emergency lights and siren. When the driver of a pursuit vehicle increases speed or drives in such a manner as to endanger the safety of others, patrols should turn on the siren and emergency lights and continuously use both throughout the pursuit. The warning effect of the siren will decrease rapidly as the speed of the pursuit vehicle increases.

   b. Radio procedures. When the pursuit begins, call the dispatcher and relay the following information:

      (1) Location, direction, and speed of travel (update continuously).

      (2) Exact reason for pursuit.

      (3) Vehicle description, including license number and number of occupants.

      (4) Traffic conditions.

      (5) Other details that would enable other patrols in the area, as well as the NSF, to assist.

         (a) Use the radio sparingly and keep the frequency open for the dispatcher and other units to assist. In the case of a two-person patrol, the rider conducts the radio communications.

         (b) While transmitting information to the dispatcher or other units, speak as normally and as coherently as possible.

         (c) When finished, be sure to place the microphone on its hook if this can be accomplished safely.

         (d) Passengers. If carrying other personnel in the car, such as prisoners, witnesses, or subjects, do not become engaged in a pursuit. Personnel involved in ride-along programs must be dropped off at a safe haven before the driver becomes involved in a pursuit.

   c. Assistance.

      (1) Patrols responding to assist in the pursuit should concentrate on covering the streets parallel to the one the pursuit is on, thus creating a boxing-in effect that may discourage the violator from continuing his/her flight. There should never be more than two police vehicles assisting in a pursuit.

      (2) This technique is also advantageous in the event the violator is able to get away from the immediate pursuit vehicle or in case the violator abandons the vehicle to flee on foot. If the violator should abandon the vehicle and flee on foot:

         (a) Remove the patrol vehicle ignition keys.

         (b) Quickly check the violator's vehicle for occupants who may have hidden in it.

         (c) To the fullest extent possible, report to the dispatcher the location, a description of the car and its occupants, and the subject's direction of travel.

   d. Maintaining a safe distance. During pursuit, a safe distance (4 seconds' following distance) should be kept between both cars, enabling the duplication of any sudden turns and lessening the possibility of a collision in the event of a sudden stop.

e. Safety belts. The use of safety belts in the patrol vehicle is mandatory. Likewise, clipboards, flashlight, and other loose objects lying in the car can become projectiles during a sudden stop, so keep them safely secured.

f. Potentially dangerous situations. Because of the potential dangers involved, NSF personnel should not pull alongside a fleeing motorist in an attempt to force the subject into a ditch, curb, parked car, or other obstacle.

    (1) Never pass a violator while in pursuit. The danger of an accident is increased, and the opportunity for escape becomes greater through quick application of the brakes and a sudden turn by the violator.

    (2) To avoid being apprehended, many motorists take dangerous chances. Regardless of the extenuating circumstances, NSF should not duplicate any hazardous maneuvers.

    (3) In the apprehension of traffic offenders and other violators, a patrol must be sensitive to safety. This means operating the vehicle in a manner that shows consideration for:

        (a) The patrol officer's safety.

        (b) The safety of the violator whose apprehension is sought.

        (c) Above all, the safety of others using the same roadway.

    (4) NSF personnel must recognize and accept the fact that one will not be able to successfully apprehend every perpetrator who flees from them.

g. Use of firearms. Refer to Appendix B of this NTTP and local policy.

h. Use of roadblocks. Because of the extreme and obvious dangers inherent in the use of roadblocks in pursuit situations, setting up roadblocks for the purpose of apprehending wanted suspects must not be used when it is likely that innocent persons would be endangered.

i. Terminating a pursuit.

    (1) NSF personnel must use their best judgment in evaluating the pursuit and continuously consider whether to continue the pursuit. Never let a personal challenge enter into the decision. The patrol should be aware the decision to abandon pursuit is, under certain circumstances, the most intelligent and most professional course of action. Stop any pursuit when the hazards of exposing the NSF member or the public to unnecessary dangers are high or the environmental conditions show the futility of continued pursuit.

    (2) It is difficult to describe exactly how a fleeing motorist could or should be apprehended except that it must be done legally and safely. It is also difficult to list any particular traffic regulations personnel could or should not disregard. Likewise, one cannot set a safe, maximum pursuit speed. NSF members must use their own judgment, their training and overall experience, and guidelines in this publication, and apply them collectively to the existing circumstances. The WC should monitor and may terminate all pursuits.

C.9 TRANSPORTING APPREHENDED/DETAINED PERSONNEL

The safety of the NSF member must be ensured when transporting persons in custody.

    1. NSF shall use the minimum force necessary to apprehend, detain, transport, and process violators.

2. NSF shall restrain and search all apprehended personnel. When restraining any subject with handcuffs or other devices, the hands shall be behind the back unless approved travel cuffs/waist chains are used. Handcuffs shall be double-locked.

3. When transporting prisoners in vehicles, NSF shall:

    a. Secure prisoners in the vehicle by use of seatbelts. Offenders shall not be handcuffed to any part of the vehicle.

    b. Not engage in vehicle pursuits, high speed, or erratic driving.

    c. Inspect the prisoner compartments of vehicles for contraband and weapons prior to and after each use and search prisoners prior to placing them into a vehicle for transport.

4. When prisoners or subjects are to be transported by a person of the opposite gender, the driver shall notify the dispatcher of the vehicle's mileage and the time before starting the transport. A record of the time and mileage shall be recorded in the desk journal. Upon arrival at the destination, the patrol shall notify the dispatcher again of the ending mileage and time. The time and mileage shall be recorded in the desk journal. When possible, another patrol officer should ride in the transporting vehicle or follow in a vehicle directly behind.

C.10 RESPONSE PROCEDURES

NSF respond to many types of incidents. Regardless of the type of emergency, operate the vehicle with extreme caution. Driving under emergency conditions does not relieve drivers of the responsibility to drive with due regard for the safety of all persons, nor will these provisions protect the driver from consequences of his/her disregard for the safety of others. Respond to an incident as Code 1 (Routine), Code 2 (Urgent), or Code 3 (Emergency). Most often the dispatcher will direct the response code. However, situations will occur requiring the NSF member to make the proper response decision.

1. Code 1: Respond by observing all applicable traffic laws. Never use emergency lights or siren for any routine call. If the vehicle operator becomes aware of circumstances unknown to the dispatching agency, the operator may upgrade the response to Code 2 or Code 3.

2. Code 2: A call requiring an immediate response to a nonlife-threatening emergency is normally assigned a Code 2 or "urgent" priority. Respond by observing all applicable traffic laws. Use emergency lights for all urgent calls. Sirens are not authorized.

3. Code 3: A call requiring an immediate response to a life-threatening emergency or in response to an emergency involving Navy high-priority resources is normally assigned an "emergency" or Code 3 priority. The use of emergency lights and siren is normally mandatory; however, use common sense when approaching the scene of the emergency. If the emergency lights and siren put NSF, victims, or bystanders in peril, turn them off at a safe distance from the scene.

4. Code 4 is used by dispatchers and NSF personnel to notify all concerned that the situation is under control.

    When responding to any incident, the patrol officers should mentally prepare themselves to perform the necessary duties. The primary functions at the scene are to control the situation, help victims, and maintain communications with the dispatcher and other patrols. Upon arrival at an incident, ensure the patrol vehicle is properly parked. Avoid having the patrol vehicle blocked by debris, other vehicles, or on-scene obstructions. There may be additional units (fire or ambulance) en route; thus, do not block the entrance to the scene. Once the scene is secured:

    a. Assess the situation.

    b. Neutralize hostile situations.

c. Attend to any injured.

d. Keep the control center and other patrols informed of the status of the situation.

e. Identify backup requirements (NSF, fire, and ambulance).

f. Identify witnesses, separate them, and advise them to remain at the scene for interviews.

g. Protect and process the crime or accident scene.

C.11 FUNDS ESCORT PROCEDURES

Unless more specific measures are prescribed by higher authorities, funds, including cash and readily negotiable instruments, shall be protected in a manner that is clearly appropriate for the amount of money involved. Armed money escorts shall not be sent off base without approval from the local authorities and the CO. While conducting escorts, NSF personnel performing such escort shall not handle the funds. Transportation for the courier shall be provided by the requesting agency. NSF personnel shall remain in a separate vehicle. SOs should establish local policy on how money escorts will be conducted aboard the installation.

1. Arming escorts. Personnel performing funds escort duties shall be armed. Those personnel armed for the purpose of escorting funds must comply with Appendix B of this NTTP.

2. Escorts and fund carriers. The localized RRP will establish procedures to detail the manner in which escorts and fund carriers operate both on and off base. These procedures should cover the positioning of the escort with relation to the fund carrier while in vehicles or on foot and the action(s) to take in the event of robbery.

3. Escort procedures. The RRP, in concert with the SO, shall establish government fund escort procedures. Formalize these procedures in the RRP. Tailor the procedures to the local threat, but address the following topics in complete detail:

   a. Procedures at the requesting activity.

   b. Frequent daytime deposits to prevent large cash buildups.

      (1) Vary deposit times.

      (2) Address fund storage limits.

      (3) Particularly address those activities that repeatedly exceed fund storage limits and routinely make deposits at closing, as this type of steady routine creates an easy opportunity for theft.

      (4) Alternative procedures in case the NSF is unable to provide the required escort (e.g., contract armored car service).

   c. Fund courier procedures.

      (1) Establish identification and duress procedures with NSF in advance of fund movement.

      (2) Drive a separate vehicle. Vary route, time of day, and, if possible, approach to depository.

      (3) Understand NSF role. NSF do not:

         (a) Carry funds containers.

         (b) Provide transportation.

(c) Have access to funds.

(4) Establish procedures for off-base movement of funds. Coordinate with civilian LE (NSF has no authority off base) and off-base depository. Recommend armored car service.

d. NSF procedures.

(1) Whenever possible, augment high-value escorts with one armed patrol or one MWD patrol, as determined by the SO.

(2) Ensure NSF escort vehicles have all emergency equipment (lights, siren, PA system, two-way radio, etc.).

(3) Obey traffic codes.

(4) Use lights and siren only for emergencies.

(5) Follow courier vehicle in a manner to preclude the possibility of being stopped or trapped together while maintaining constant visual observation (often this is nothing more than prudent vehicle separation for the posted speed limit).

e. Communications procedures.

(1) Coordinate escort itinerary (time and place of departure, route, destination, and estimated time of arrival).

(2) Security checkpoints (required at start, periodically, and upon successful completion of the deposit).

(3) Duress alternatives.

f. Communication outage procedures (loss of communications with a high-value fund escort should result in antirobbery procedures initiation until status of escort can be determined)

g. Development of backup emergency response procedures.

C.12 BUILDING SECURITY CHECKS

One of the NSF responsibilities under the installation security plan is to make security checks of weapons and munitions storage areas, pharmaceutical and supply repositories, and other buildings and areas that are secured. Conducting building checks is an excellent form of proactive crime prevention.

1. Each activity shall establish a system for the daily after-hour checks of restricted areas, facilities, containers, and barrier or building ingress and egress points to detect any deficiencies or violations of security standards.

2. All building security checks shall be documented in the desk journal, and security violations must be reported to the SO.

a. Records of security violations detected by NSF personnel shall be maintained for a period of three years.

b. The SO must follow up each deficiency or violation and keep a record of all actions taken (structural, security, disciplinary, administrative, etc.) to resolve the deficiency or violation and how further recurrences will be prevented.

3. Below are a few reasons and points to keep in mind when conducting building checks:

    a. Visibility of NSF patrols to the Navy community.

    b. A public relations opportunity to meet other military and civilian workers in their environment and provide them assurance that their professional property will be protected when they are away from their duty station.

    c. Keeps vandals and would-be thieves at bay, uncertain as to when and where NSF patrols might arrive.

    d. Provides LE patrols an opportunity to learn building layouts, likely avenues of approach/escape, and safe and efficient response routes. This benefit not only applies to nonduty hours but will improve response capabilities and enhance officer safety when answering calls for assistance during duty hours.

    e. Be familiar with the kind of work performed in each facility, potential hazards of stored materials (if any), and the exact location of resources in the building. Take time to learn as much as possible about the facilities on the installation. Meet the building custodians and discuss their concerns. This information may be extremely valuable during an incident at the facility. Responding to a crime in progress is not the time to learn about a building.

C.13 BUILDING SECURITY CHECK PROCEDURES

1. Desk journal entries shall include the facility number, start and stop time of the check, the patrol conducting the check, and the results of the security check.

2. Personnel may check the same buildings and areas each day, so it is important that NSF personnel not set a predictable pattern. Approach the facility from a different direction each time, and do not check the same structures at the same time each day. If the routes and times are predictable, someone can easily avoid being caught in the act of breaking in or burglarizing. As patrol officers approach, they should be alert for suspicious vehicles or activity. Notify the dispatcher when the patrol arrives, before start of the security check, and once checks are completed.

    a. As patrols approach the building, try to stay out of well-lit areas and stay alert for suspicious activity. Fire escapes, rooftops, and buildings constructed off the ground on stilts provide a perfect hiding place for an intruder to gain access to a building.

    b. When conducting building checks, look for obvious signs of forced entry, such as broken windows, pry marks, or open doors.

    c. Physically check all entrances to the building that the NSF member can reach. Try to open doors and windows (within reason) and turn all door handles.

    d. Look closely for signs of forced entry. When possible, attempt to look inside buildings through windows.

    e. Remain unpredictable and do not set a pattern.

3. If a patrol finds an open window, broken window, or any sign of forced entry:

    a. Take cover immediately from a position where the facility can be observed but where the observer cannot be seen and is not silhouetted.

    b. Once in a covered position, contact dispatch and inform them of the incident, building number, patrol's location, and any other important facts.

    c. The dispatcher shall immediately dispatch the appropriate backup patrol(s) and contact the building custodian. Do not enter the building until backup patrols arrive. Once backup arrives, coordinate a plan of action with the on-scene patrols and the dispatcher.

d. If available, use MWD teams to search and clear an unsecured building.

e. Once the building point of contact arrives on scene, NSF search the building, apprehend any unauthorized persons, and remove them from the area.

f. Upon completion of the building search, the point of contact checks to see whether theft or vandalism has taken place or someone merely forgot to secure the building.

g. Initiate an IR and obtain all pertinent statements, investigative notes, and any other required reports.

INTENTIONALLY BLANK

APPENDIX D

External Entry Control and Restricted Area Access Control

D.1 GENERAL

ECPs are only one aspect of the installation PS efforts, but they are extremely important to defense-in-depth and effective risk mitigation. The primary purpose of the ECP is to provide positive access control at the perimeter. It should be designed and manned to prevent unauthorized access and intercept contraband while maximizing authorized access to an installation or area.

D.2 ENTRY CONTROL POINT OPERATIONS

Identification and inspection are the most common operations conducted at an ECP. The level of identification and inspection varies with the force protection condition (FPCON). At all levels of FPCON, identification procedures shall be adequate to ensure the vehicle occupants are authorized access.

The installation AT plan shall define the operations of the ECP during each FPCON and shall include RAM from each FPCON at any time.

The design and operation of ECP should support FPCON Bravo operations with minimal congestion.

D.3 STRUCTURE OF AN ENTRY CONTROL POINT

An ECP can be subdivided into four zones, each encompassing specific functions and operations. Beginning at the installation property boundary, the zones include the Approach Zone, Access Control Zone, Response Zone, and Safety Zone. Specific components are used within each zone to conduct the necessary operations. The location of each zone of the ECP is illustrated in figure D-1.

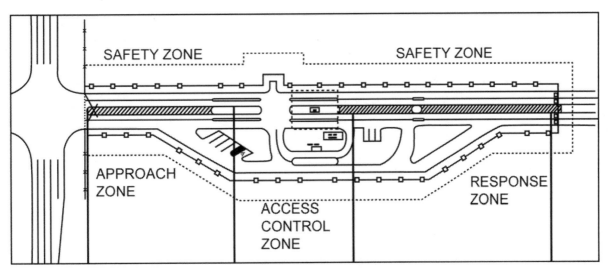

Figure D-1. Entry Control Point Zones

1. The Approach Zone lies between the installation boundary and the access control zone. It is the interface between the off-installation road network and the installation and the area all vehicles must traverse before reaching the actual checkpoint. The approach zone should support the following functions and operations:

    a. Reduce the speed of incoming vehicles to, or below, the designated speed of the ECP

    b. Perform sorting of traffic by vehicle type (e.g., sorting trucks or visitors into the proper lane before reaching the inspection area or checkpoint)

    c. Provide adequate stacking distance for vehicles waiting for entry, especially during times of peak demand, to ensure minimal impact on traffic approaching the installation and on traffic safety operations of adjacent public highways

    d. Provide the first opportunity to identify potential threat vehicles, including those attempting entry through the outbound lanes of traffic.

    Roadway layout and traffic control devices such as signs, variable message systems, signals, and lane control markings should be utilized to perform these functions. Drivers should be notified of the upcoming access control point, the proper speed to travel, and proper lane to utilize. The length of the approach zone is based on available land, distance required for queuing and performing traffic sorting, and the space required to create additional lanes of traffic without queuing excessively onto adjacent public highways. The ECP should also support measures that may be needed during higher FPCON levels, the use of RAMS at lower FPCON levels, and the temporary placement of traffic barriers as specified in the Installation AT Plan to constrain and slow traffic. Space may also be required to support traffic calming techniques to mitigate high-speed threats.

2. The Access Control Zone is the main body of the ECP and includes guard facilities and traffic management equipment used by the NSF. The ECP should be able to process the following types of vehicles depending on the intended functions of the ECP:

    a. POV of authorized personnel

    b. Government vehicles

    c. Visitor vehicles

    d. Military convoys

    e. Delivery vans, trucks, and buses.

3. Most installations conduct identification procedures manually and require both vehicle and personnel identification. NSF generally conducts these procedures in access control zones at FPCON Bravo and below:

    a. Verification of vehicle decals

    b. Verification of personnel identification

    c. General surveillance of the vehicle and its contents

    d. Random, complete inspections of the vehicle and contents.

    Most installations issue visitor and/or vehicle passes at a centralized visitor's center. If a vehicle is denied entry during identification checks, the access control zone must have room for that vehicle to be redirected to exit the installation. Traffic arms can be used to control traffic when a vehicle is being rejected from the

ECP. Installations may use tandem processing, with two or more security personnel posted to each lane of traffic, to increase the throughput of an ECP. It has been estimated that tandem processing may improve capacity by up to 50 percent per lane. This additional capacity may be critical during increased FPCON levels or during the use of RAM.

4. The Response Zone is the area extending from the end of the access control zone to the final denial barrier. This zone defines the end of the ECP. The response zone should be designed so that NSF has time to react to a threat, operate the final denial barriers, and close the ECP if necessary.

5. Safety Zone. The safety zone extends from the passive and active barriers in all directions to protect installation personnel from an explosion at the vehicle barricade. A terrorist vehicle could explode inside the contained area of the ECP. The size of the safety zone is determined by the acceptable standoff distance by the expected weight of the explosive charge and the facility or asset to be protected. If an adequate safety zone or standoff distance cannot be achieved to produce acceptable damage and injury levels, evaluate other alternatives or the decision must be made to accept additional risk.

D.4 INSTALLATION AND RESTRICTED AREA ACCESS CONTROL

Each commander or tenant shall:

1. Clearly define the access control measures required to safeguard facilities and ensure accomplishment of the mission. These measures shall be identified in installation security plans and shall be included in applicable departmental RRP. Security plans shall include:

    a. A defense-in-depth concept to provide graduated levels of protection from the installation or activity perimeter to critical assets.

    b. Positive access control measures at the ECPs.

    c. The degree of control required over personnel and equipment entering or leaving any exterior control point and any restricted area enclaves within it, including a description of access control measures in use and the method for establishing authorization for entering and leaving each area as they apply both to personnel continually authorized access to the area and to visitors, including any special provisions concerning nonduty hours.

    d. Use of security badges and military identification (ID) cards.

    e. Procedures for inspecting persons, their property, and vehicles at entry and exit points at any exterior control points and restricted areas.

    f. Use of RAM within existing security operations to reduce patterns, change schedules, and visibly enhance the security profile.

    g. Process for coordinating with local, state, federal, or HN officials, as well as tenant organizations, to ensure integrity of restricted access and reduce the effect on primary missions and surrounding civilian communities.

    h. Maintenance of adequate physical barriers that shall be installed to control access to the installation/activity or restricted area.

    i. Process for removal of or denying access to persons who are not authorized or are a threat to order, security, and discipline.

    j. A mechanism to keep appropriate personnel informed of the plan and their responsibilities.

    k. Maps where additional bollards are located on the installation and responsibility for transporting to a location if needed.

    l. Hours of operation and required manning levels during each FPCON for all ECPs.

2. Ensure that access control measures are IAW OPNAVINST F3300.53 (series), Navy Antiterrorism (AT) Program. At FPCONs NORMAL, ALPHA and BRAVO:

   a. Additional unarmed personnel may be assigned as ID checkers to maintain smooth traffic flow at all installation perimeter vehicle ECPs.

   b. While in home port on normal workdays or during other periods of heavy traffic, ships/squadrons will augment their respective pier/flight line ECP with at least one additional contact sentry to support visitor control and traffic flow. At those times when pier/airfield ECPs are manned by a single sentry, vehicle ECPs will remain closed except when opened for vehicle passage.

D.5 AUTOMATED ENTRY CONTROL SYSTEMS

Detailed guidance pertaining to Access Control Systems can be found in UFC 4-021-02NF, Security Engineering: Electronic Security Systems and UFC 4-022-01, Security Engineering: Entry Control Facilities/Access Control Points. Automated entry control systems include the following:

1. Electronically operated gates to be activated by security personnel at the ECP, from a dispatch center, or by a card/badge reader using either contact or (preferably) contactless technology.

2. Closed-circuit television (CCTV) with the capability to display full-facial features of a driver or pedestrian and vehicle characteristics on the monitor at the security office.

3. An intercom system located in a convenient location for a driver/pedestrian to communicate with the ECP sentry or security office.

4. Bollards or other elements to protect the security booth and gates against car crash

5. Sensors to activate the gate, detect vehicles approaching and departing the gate, activate a CCTV monitor displaying the gate, and sound an audio alert in the installation dispatch center or security office.

6. Signs to instruct visitors and employees.

D.6 ENTRY CONTROL POINT PROCEDURES

D.6.1 Entry Control Point Threat

1. Any person or vehicle that needs to reach a critical asset or area should be required to pass through an ECP. At Navy installations and commands, ECPs are typically base gates, pier accesses, and ships' quarterdecks. Such defense-in-depth is designed to keep pedestrian-carried and vehicle-borne improvised explosive devices (IEDs) far enough from critical assets and areas to prevent serious damage.

2. Procedures for getting onto an installation may differ from those procedures required to gain entry to a pier or ship. Terrorists will likely be familiar with ECP procedures and will tailor their actions to appear inconspicuous. (Unless suicidal, terrorists do not want to advertise their intentions.) The slightest oddity on an ID card or mannerism may be all a terrorist reveals. When developing RRP for establishing and maintaining an effective ECP, it is important to do the following:

   a. Analyze the latest threat assessment.

   b. Conduct a prearrival pier sweep to establish a landward defensive zone.

   c. Review HN limitations (e.g., number of U.S. sentries, types of weapons, HN versus U.S. responsibilities).

d. Ensure contact information for HN, explosive ordnance disposal (EOD), police, and emergency personnel is current and available.

e. Identify likely avenues of approach for threats.

f. Provide multiple means of communication between the sentries and the protected asset.

g. Post signs to establish and define the security perimeter and ECP that are visible from a maximum distance, lighted at night, and carrying warnings, written in the local language and English, to personnel to remain clear of the restricted area.

h. Identify ECP manning requirements, including contact sentries, inspectors, and cover sentries.

i. Identify cover positions for sentries.

j. Identify positions for cover sentries and types of weapons.

k. Identify minimum evacuation distances if an IED is discovered.

l. Provide the sentries with important phrases in the native tongue.

m. Equip watch standers with weapons, personal protective equipment (PPE), body armor, and protective masks.

n. Ensure communications, challenge procedures, and inspections procedures are well-rehearsed.

o. Coordinate warning procedures in preparation for an incident. Warning procedures can include radio, pyrotechnics, hailers, whistles, loudspeaker system, mirror, and hand and arm signals.

p. Identify lighting placement around ECP (lighting plan).

q. Identify detection methods and measures to be used, including MWD teams, explosive detection devices, chemical agent detectors.

r. Determine who and what will be inspected and to what degree.

s. Ensure baggage and package inspectors are aware of the following suspicious indicators:

    (1) Weight: unevenly distributed, heavier than usual for its size, heavier than usual for its postal class.

    (2) Stamps and postmark: more than enough postage, foreign city.

    (3) Thickness.

        (a) For medium-size envelopes, the thickness of a small book.

        (b) Not uniform or has bulges.

        (c) For large envelopes, an inch or more in thickness.

    (4) Writing: foreign writing style; misspelled words; marked "air mail," "registered," "certified," or "special delivery."

(5) Address: marked "personal," "private," or "eyes only"; no return address; poorly typed or handwritten address; hand-printed; incorrect title for recipient; addressed to a high-ranking recipient by name, title or department.

(6) Envelope: peculiar odor; oil stains; inner-sealed enclosure; excessive sealing material; wire, string, or foil sticking out; ink stains.

(7) Rigidity: springiness; greater than normal, particularly along the center length of the package/baggage.

t. Identify location for personnel and package inspection area.

u. Sandbag the inspection area to minimize blast damage.

v. Request HN/base security inspect buildings and clear away abandoned vehicles surrounding an ECP on a continuous and random basis to deter surveillance.

w. Establish a vendor access list with a schedule of all deliveries.

x. Rehearse crowd control procedures and coordinate these with HN or local authorities.

y. Deny access to obviously ill personnel and refer them to local health care providers.

z. Identify fields of fire for crew-served weapons, and minimize potential collateral damage while maximizing coverage of ECP.

aa. Develop barrier plan to channelize threats.

bb. Consider establishing additional or secondary ECPs closer to the protected asset.

cc. Conduct guard mount and post sentries as required.

D.6.2 Personnel and Vehicle Identification

Before allowing personnel and vehicles to pass through an ECP, watch standers shall follow the procedures below. In FPCONs NORMAL, ALPHA and BRAVO, installations may incorporate a Trusted Traveler Program as local security conditions permit that allow an individual with authorized access to a Navy installation to vouch for and accept responsibility for all vehicle occupants.

1. Check ID cards and ensure at least one individual in the vehicle is authorized access to the installation or check all personnel if required by installation or higher headquarters policy.

2. Conduct a visual and physical inspection of the identification card(s) and visually match the photograph on the card to the person presenting the card.

3. Ensure ID card is authentic and the identification document is approved for installation access.

4. Confiscate altered ID cards, retain person, and turn over to commander of the guard (COG)/patrol supervisor or ECP supervisor.

5. Inspect personnel and hand-carried items per current policy.

6. Ensure bearer is authorized entry per current access policy or daily access list.

7. Detain unauthorized personnel attempting to gain access and notify the COG/patrol supervisor or ECP supervisor.

8. Check vehicle ID papers if required by installation or higher headquarters policy.

9. Direct vehicles identified for inspection per local policy to the inspection area and inspect per current policy.

10. Deny access to unauthorized vehicles.

D.6.3 Vehicle Inspections

1. NSF should make use of entry- and exit-point vehicle inspections to ensure compliance with vehicle safety and registration requirements, safeguard government property, control contraband, and deter and detect impaired driving. Mobile vehicle inspection teams shall be used both at ECPs and within the installation to conduct vehicle inspections as part of the installation PS program. Mobile vehicle inspection teams shall consist of at least two inspectors and a supervisor. Persons and vehicles attempting to enter an activity may not be inspected over the objection of the individual; however, those who refuse to permit inspection shall not be allowed to enter. Persons who enter shall be advised in advance that they and their vehicles are subject to inspection while aboard the activity or within the restricted area. (A properly worded sign to this effect prominently displayed in front of the entry point will suffice.) There are three general categories of vehicle inspections conducted by ECP and mobile inspection team personnel.

   a. Administrative inspection. A vehicle inspection that includes the verification of occupant identity and, in the case of commercial vehicles, the verification of delivery documents (e.g., bill of lading). These tasks may be satisfied by official access lists and other accepted documents.

   b. Intermediate inspection of vehicles. As in administrative inspections, intermediate inspections must verify driver identity, cargo, and destination. Additionally, intermediate inspections are done to observe the entire vehicle exterior portions in plain sight, including under the vehicle. Limited questioning of drivers should take place during intermediate inspections. This level of inspection is used as a part of spot checks.

   c. Complex inspection of vehicles. In addition to the requirements of the administrative and intermediate inspections, complex inspections are done to observe all exterior and interior vehicle spaces normally accessible without the use of special tools or destruction of the vehicle infrastructure. Nondestructive inspection of vehicle cargo and passenger areas is expected. This procedure includes physical observation of vehicle interior, including cargo spaces that may be sealed.

2. Personnel responsible for the accomplishment or implementation of personnel and vehicle control procedures shall be watchful for the unauthorized introduction to or removal from the installation of government property, especially weapons and AA&E materials. This responsibility includes all personnel and means of transportation, including government, private, and commercial vehicles, aircraft, railcars, and ships.

Paragraphs D.6.3.1 through D.6.3.9 contain basic guidelines for NSF who conduct access control and perform complex inspections of private and commercial vehicles for contraband and explosives. The inspection of every vehicle is unique. The guidance below is to be applied in conjunction with previous training, experience, and standard procedures and policies. Be aware that each case presents its own unique circumstances. In all cases, use common sense and do not conduct any vehicle inspection that would place personnel, equipment, or facilities at risk. Included are techniques for interviewing personnel and inspecting various types of vehicles commonly found at installation ECPs and commercial vehicle inspection facilities. Inspecting vehicles and interviewing drivers/passengers are critical tasks when providing installation PS. A single vehicle carrying explosives and/or explosive devices can cause massive damage and bodily harm. Installations should establish standardized interview and inspection procedures based on the guidance provided in this section to ensure a thorough inspection and avoid duplication of effort. The important thing in any inspection is not to miss any area that normally would be inspected. This section should not be considered a complete source for personnel interviews and vehicle inspections. Use these guidelines along with other available resources, such as the Technical Support

Working Group Vehicle Inspection Guide, Regional and COM-specific guidelines, and personal experience, to develop installation-specific techniques.

D.6.3.1 Driver Interview

The vehicle inspection interview process is used to gather information on the vehicle driver and passengers (if any). How this process is conducted will determine the probability of identifying an attempt to introduce contraband or explosives into a Navy installation. Be aware of the cultural customs in the area. This may affect the person's manner toward NSF, particularly with different genders. Separate the behavior that is cultural from that which is stress-related. Remember that the interview process is not an exact science. An innocent driver could demonstrate physical indicators, while a guilty driver could have no physical indicators. Don't overcomplicate the interview. Ask questions until satisfied with the responses, make an assessment, confer with the other inspectors, conduct the vehicle inspection if required, and then hold or release the vehicle. If the decision is made to inspect a vehicle, the driver and all occupants should exit the vehicle, stand a reasonable distance away from the vehicle and inspectors during the inspection and remain under the observation of an NSF during the entire procedure.

D.6.3.2 Obtain Knowledge of the Typical Entry-Point Person and Vehicle

1. What vehicle types have been identified as potential threats in the latest intelligence briefings?

2. What types of people have been identified as potential threats? Do any of these drivers and vehicles fit these profiles?

3. What type of person do you normally encounter at the ECP at this particular time?

4. How does the average person of this age and social status dress when entering the installation for a particular purpose?

5. What are the normal amounts and condition of the cargo of the vehicle?

6. What is the normal type of vehicle entering the facility at this time of day?

7. What are the common reasons given for a specific type of vehicle to be entering the installation?

8. What are the typical types of vehicles driven by persons of this particular age group, dress, occupation, and social status?

9. What are the typical types of passengers for this type of vehicle?

10. Do the identification documents match the person being interviewed?

11. What work groups are scheduled at what times throughout the day?

12. What are the typical means of transportation for each work group?

D.6.3.3 Initial Observation

Make an initial observation of the vehicle/driver entering the inspection area. Does any area catch the eye as being out of the ordinary? Examples to look for:

1. Visible fingerprints or smudges around the front bumper, headlamps, fender wells.

2. License plate is clean, but the vehicle is dirty.

3. Tires are low, too clean for the vehicle, do not match.

4. Vehicle looks overly heavy in either the front or the rear.

5. Driver appears unfamiliar with the clutch or brakes.

6. Occupants fail to fit the vehicle:

   a. Sloppy-appearing person in a neat automobile.

   b. Neatly dressed person in a messy or very dirty vehicle.

   c. Occupants appear to be tense, overly friendly, too casual.

D.6.3.4 Conducting the Interview

1. The following list contains typical questions to obtain driver knowledge specific to the vehicle type. Ask in random order, but not all questions need to be asked (* = commercial).

   a. What is your citizenship?

   b. Where do you live?

   c. Where are you going?

   d. What are the name and position of the person you are to see?

   e. What is your cargo? (*)

   f. Where did you come from?

   g. By whom are you employed?

   h. Where is your place of work?

   i. Do you own the vehicle? If not, who owns it? Do you know the owner?

   j. How long have you been employed by the present employer? (*)

   k. Do you drive this vehicle most of the time?

   l. How long do you expect to be on the installation?

   m. Did you observe this vehicle being loaded? (*)

   n. May I see your logbook? (*) (U.S. only)

   o. Would you be aware of any contraband in the vehicle?

   p. Has the vehicle been worked on/repaired recently? If yes, what was repaired?

   q. Additional questions specific to the vehicle type.

   r. Where does the owner live?

2. Indicators from the interview that may indicate further investigation or inspection is required:

   a. Driver inappropriately dressed/groomed for vehicle type.

    b. Driver operating a commercial vehicle does not have a commercial driver's license.

    c. Driver does not have a logbook. (*) (U.S. only)

    d. Driver's story does not match documentation.

    e. Documentation incomplete, does not make sense.

    f. Driver does not know his or her purpose and/or destination or documentation.

    g. Driver and passenger(s) do not appear to be on the same mission.

    h. Driver does not know how to operate vehicle or equipment.

    i. Presence or lack of significant odors consistent with cargo.

3. If necessary, have the driver exit the vehicle and ask him/her to open any baggage, container, or vehicle. Do not lead the way—follow the driver. The driver should open all compartments or baggage unless physically unable to do so.

4. Observe the person opening the compartment(s) and baggage.

    a. Is the person familiar with the vehicle?

    b. Does the person hesitate or appear nervous?

    c. Do the person's hands shake while opening the compartment?

    d. Do the person's eyes dart back and forth?

D.6.3.5 Stress Indicators

1. Stress is an uncontrollable instinct. The manifestations of stress are initiated instinctively. This means that when a threat to a person's well-being is perceived, an automatic, uncontrollable animal instinct occurs. Physical and psychological changes occur immediately, produced by the autonomic nervous system present in all animals. The common term for this process is the fight-or-flight mechanism. Upon the realization of an impending threat to one's well-being, the body automatically prepares to fight or flee the danger. Watch for these physical manifestations of the fight-or-flight mechanism:

    a. Turning red or blushing

    b. Turning white or pale

    c. Obvious shaking

    d. Averting eyes, refusing eye contact

    e. Darting of eyes

    f. Evasive eyes or looking at the floor

    g. Patting the cheek (soothing and reassuring gestures)

    h. Yawning (if excessive or repeated, this is a very strong indicator of deception)

i. Covering the eyes

j. Tugging at clothing or any area of the body

k. Tapping the chest

l. Wiping hands

m. Sitting on edge of seat (preparing for flight)

n. Rubbing hands or fingers together

o. Licking the lips

p. Biting or chewing the lips

q. Twisting the mouth

r. Covering the mouth

s. Protecting throat area with the hands

t. Swallowing repeatedly/excessively

u. Pulsating carotid artery

v. Sweating profusely when the environment, dress, or activity does not warrant it

w. Exhibiting goose bumps

x. Hairs standing on end of the arms and back of the neck

y. Fidgeting/nervous hands

z. Playing with fingernails

aa. Patting/soothing/stroking/massaging any area of the body (reassuring gestures)

bb. Scratching repeatedly

cc. Dilating pupils.

2. Listen for auditory manifestations of the fight-or-flight mechanism:

   a. Inability or reluctance to answer the question

   b. Answering a question with a question

   c. Repeating the question or asking for repetition

   d. Continually asking for clarification

   e. Not answering the question

   f. Responding with unrelated information (rehearsed answers)

    g. Hesitant speech

    h. Deep sighing

    i. Repeatedly clearing throat

    j. Yawning (important key)

    k. Grinding teeth

    l. Oral clicking sound (dry mouth)

    m. Attempting to influence the questioner by using the following words or qualifications: honestly, truthfully, believe me, to tell the truth, to be perfectly frank, I swear, may God strike me dead, I wouldn't lie to you.

D.6.3.6 General Guidelines for the Inspection of Passenger Vehicles

1. Look for anything in factory-built compartments. These compartments can be excellent areas for hiding items.

2. Look for new or shiny bolts and/or screws, indicating that something has been altered, modified, or moved. Shiny bolts or screws indicate that the bolts or screws are new, whereas a scratched bolt or screw indicates removal and reinsertion.

3. Look for unusual scratches or other signs of tampering, which indicate that the area possibly has been modified and the repair or replacement completed sloppily with some kind of tool. Look at sheet metal for scratches and small dents.

4. Check for unusually clean or dirty components in all areas. Clean components in an otherwise dirty area indicate that these items are new or recently repaired or modified. The driver or owner of the vehicle should be able to explain why new or clean components are in the vehicle. Be especially aware of clean or new wiring, which can be directly connected to an explosive device.

5. Check for spools or remnants of electrical wire, tape, or similar items. These materials are widely used in making vehicle bombs and may indicate that some type of device was constructed in the vehicle.

6. Look for new or broken welds. A common technique in hiding items in deep concealment is to cover them with metal welded to the vehicle. Sometimes, existing welds are broken and rewelded to hide devices.

7. Look for unusual dirty or greasy fingerprints on exterior surfaces. When someone is working in an area, handprints and smudges can be left behind where road dirt has accumulated.

8. Check the headlight/taillight wires. The wiring for the lights can be used to connect an electrical source to a detonator.

9. Look for any fresh bodywork. A close inspection of the body will reveal new bodywork. A typical way to hide devices is to create a false compartment in a section of the body. This false compartment is then covered with fiberglass, paint, undercoating, etc. (Be aware that in parts of Europe the undercoating is the consistency of grease.)

10. Feel or look at the back of the bumpers for a false compartment. The bumper is typically made of steel plate and should not be overly thick.

11. Look in the area between the front grill and the radiator. This area can be used to hold an explosive device. Electrical power from headlight wires is readily available.

12. Exterior sides:

   a. Look at the fenders for an unusually thick or wide area. False compartments can be integrated into the fenders, especially in the areas above the tires. Look for fresh bodywork and undercoating.

   b. Swing the doors. (Have the driver initially open the doors.) Be aware of the feel of a normal door. If the door feels heavy when swung, then something may be hidden inside the door. The door cavity is an excellent place for hiding materials. Inspection holes are also found on the edge of the door by the latch. By removing the rubber plug, the inside of the door can be inspected. Rolling down a window is another way to search a vehicle's doors for concealed material.

   c. Tap on the bodywork along the side of the vehicle, using your hand or a small rubber mallet. (Do not damage the vehicle.) Listen to the sounds. The length of the vehicle should sound hollow. If the sounds are inconsistent, there may be a change in density caused by material hidden inside the door(s).

13. Observe everything within view in the engine compartment, looking for anything obviously unusual.

   a. Check the battery box for anything unusual. Some vehicles have large battery boxes that can be used for a smaller battery and have enough volume to act as a cavity for explosive materials.

   b. Look for odd or additional wires running from the vehicle's battery. Look for additional wires running from the headlight or the absence of a bulb in the headlight socket. Look for any unusual wiring. The battery is an excellent source of power for a detonator, and the vehicle wiring provides an easy way to connect the battery to the detonator.

   c. Look for out-of-place or unusually clean components, devices, and/or wiring and electrical tape. Unusually clean areas, new tape, and/or new devices mean recent modifications or installations. The driver should be able to explain the clean areas and/or devices.

   d. Check the windshield washer container for a false compartment or signs of tampering.

   e. Inspect the firewall for any signs of modification or tampering. Especially look for signs of sheet metal work indicating the possibility of something hidden behind the firewall. Also look for new welds and new or shiny screws.

14. Observe everything within view inside the vehicle. Pay close attention to packages/devices (e.g., alarm clocks, iron or PVC pipe) that look out of place.

   a. Look closely at the dash for new, damaged, or scratched screws. Behind the dash is a factory-built cavity. To get access, the dash has to be removed. Removing and reinserting the screws can cause damage. A smuggler may decide to use new screws instead of reinstalling the old/damaged ones.

   b. Look for plugged vents on the dash. The vents are a primary location for hiding contraband.

   c. Look for bits of electrical tape, wire, stripped insulation, string, fine wire, fishing line, and/or a time fuse on the floor, dash, or seats. All of these materials are commonly found in explosive devices.

   d. Check under floor mats for wires or switches that could be part of an explosive system.

   e. Look at the glove box. Be aware of the typical look and depth of a glove box. Be suspicious of any glove box that looks small and/or shallow. A false compartment could be present.

   f. Check the seats, especially the rear seat, for unusual bumps or bulges. If bumps and/or bulges are present, find out why they exist.

g. Check the roof liner for bulges, rips, and/or repairs indicating possible concealment of an explosive device or explosive materials.

h. Look at the floor for anything that appears to be modified. Remove floor mats if necessary. Typically, if anything is hidden in the floor, the floor will appear overly thick. Look for fresh welds or seams.

i. Look for packages, containers, travel bags, or devices that seem out of place. An explosive device will not necessarily be hidden in the vehicle. It may be hidden in a cardboard box on or under the seat.

15. Cargo area:

a. Look for any new items in the trunk, such as carpeting or mats, indicating something has been changed recently.

b. When looking in the trunk, be aware of the smell of caulking, glue, or any other unusual materials indicating that something has recently been modified or repaired.

c. Move the trunk lid up and down. (Have the driver open the trunk for you.) Notice whether the trunk lid feels heavy or will not stay up. A heavy trunk lid indicates that something may be hidden in it.

d. Look for an unusually high or unusually thick trunk floor. A false floor can be used to hide items below it.

e. Tap on the walls of the trunk to check for false compartments. The sounds should be hollow. Any solid sounds indicate that something may be behind the wall and should be further investigated.

f. Look around the trunk area for unusual welds or seams. Any of these could hide a loaded false compartment.

g. Check the area behind the rear seat. This is an area where a false wall could be constructed.

h. Look at the spare tire area. If the spare tire is not flush with the floor surface or does not look correctly installed, look for a false bottom to the tire well.

D.6.3.7 General Guidelines for the Inspection of Commercial Vehicles

Inspectors should take extra care when inspecting commercial vehicles because of the ability to easily hide contraband such as explosives and the increased number of concealment locations. In addition to inspecting the same areas as you would with a passenger vehicle, the following areas should be inspected on commercial vehicles. This is not an all-inclusive list and experience may dictate the inspection of additional areas.

1. Each vehicle should possess documentation that provides valuable information during an inspection: vehicle registration; logbook; insurance card; shipping papers (bill of lading); HAZMAT signage (as required); and interstate inspection tags (as appropriate).

2. Look around the fifth-wheel area.

3. Check any factory-built compartments.

4. Check for any compartments in the frame rails. Frame rails are normally open; however, new compartments can be welded, so look for new welds.

5. Look for fresh bodywork. A typical way to hide contraband is to create a false compartment in a section of the body.

6. Check the vertical exhaust pipe for functioning and false compartments. Since motors can run on one exhaust, the second pipe can be used as a hiding place.

7. Look in the area between the radiator and front grill. This area can be used to hold an explosive device and is near the headlights for electrical power.

8. Look at the battery box for anything unusual. Empty battery shells can be used as false compartments.

9. Look at the fenders and bumpers for any unusually thick or wide area where a false compartment could be hidden.

10. Fuel tanks are the primary area for hiding explosives. Look for unusual welds under the holding straps. Look for the presence of split fuel tanks. Tap on the tanks and listen for inconsistent sounds. If there is something in the tanks other than fuel, you will hear different sounds on different areas of the tanks. Look for missing connections on the tank. If something other than fuel is in the tank, the fuel lines may not be hooked to the tanks. Feel the fuel tanks for temperature consistency. The temperature of the tank along the length should be consistent whether you are feeling the filled or unfilled area.

11. When looking in the engine compartment, take a moment to observe everything within view for anything obvious or unusual. Look for odd or additional wires running from the vehicle battery, and check under larger components for hidden compartments.

12. Within the driver compartment, look for anything that appears to be modified. An usually clean interior area is suspicious since trucks are used on a daily basis and not normally kept in a clean condition. Look at the floor for anything that appears modified. If the tractor has a sleeper, look under the mattress and for the presence of luggage belonging to the driver. The absence of luggage is suspicious, as is the presence of hard-sided suitcases.

13. Use a flashlight and mirror with a creeper, if possible, to inspect under the vehicle. When looking under the vehicle, look for anything taped or attached to the frame. Be suspicious of signs of recently installed hardware. The driver should be able to explain the presence of any new hardware. Look for spare or extra tanks on the vehicle that have no obvious uses.

14. When inspecting the trailer, check the floor for unusual thickness or evidence of modifications. Also look for trailer tires obviously smaller than the adjoining tires. This may be an indication the smaller tire has explosives or other contraband in them since they would not spin or get as hot.

15. Look for new or strange air tanks on the trailer. If suspicious, use the bleeder valve to check for air pressure. If there is no air pressure, the tanks may be being used to hold contraband.

D.6.3.8 General Guidelines for the Inspection of Commercial Cargo Trucks

The typical use for box-like, closed cargo vehicles is transporting all types of merchandise and commodities.

1. All company insignias should be professionally applied. Be suspicious if the company logos are applied poorly.

2. A common hiding technique is the use of false walls. A false wall can be determined by comparing the exterior length with the length of the cargo area. If they are not the same length, look for a false wall in the front of the cargo area. Tap on the walls and roof to discover false cavities. Look at the floor surface for signs of modification. Be suspicious of new or modified floor planks that may cover a hiding area below. Look for hinges or signs of tampering where the roof and walls meet. This may be an indication of hiding contraband on top of the roof.

3. When inspecting cargo in a trailer, shine a light under the cargo pallets. You should be able to see all the way to the front wall of the cargo area. Cargo typically consists of a single commodity boxed uniformly. When looking over the top of the cargo, the top surface of the load should uniform. If there are dips, there could be something hidden in that area.

4. Notice where the cargo boxes are purchased. Companies usually purchase large quantities of boxes with their logos on the boxes. Be suspicious if the cargo is boxed in U-Haul or similarly named boxes since these boxes are expensive and shipping companies tend to avoid buying them. Also look how the boxes are stacked, as a sloppily stacked cargo could mean indifference to the safety of the cargo and that it is there strictly for diversionary reasons.

D.6.3.9 General Guidelines for the Inspection of Specialty Commercial Cargo Trucks

In addition to the inspection techniques discussed above, specialty cargo trucks require different techniques to inspect.

1. Dump trucks—Look closely for poor welds and signs of tampering or modifications to the sides of the bed and the tailgate. With a metal rod, probe into the cargo material to feel for any foreign objects. Typically, foreign objects will be on the bed surface or between layers of cargo material.

2. Garbage trucks—Have the operator or driver explain how the truck operates. If the operator does not know, he or she may have another mission.

3. Dumpster garbage trucks—When entering an installation, the dumpster should be empty. Verify it is. If not, ask the driver to explain why. A full or partially loaded dumpster can hide large quantities of explosives.

4. Refrigeration truck or trailer—The cargo should be appropriate for the trailer, i.e., it should be perishable.

5. Gasoline trailer—The potentially dangerous commodity carried by gasoline tankers requires extra caution during any inspection. Four factory-built internal compartments are normally found within the gasoline tanker, with varying capacities in each compartment. Internal baffles with manhole passageways are built into the bulkhead walls of the compartments. The highly volatile nature of gasoline and other petroleum products tends to inhibit close inspection of these tankers. In this regard, it should be noted that empty tankers are often more dangerous than those that are full due to the explosive potential of fumes in a confined area. Feel the sides of the tanker with your hand. The tanker surface temperature should feel consistent along the length of the tank. The lower part of the tank, depending on how full the tank is, should be cool, while the tank surface above the liquid level should feel warmer. Any temperature differences along the length of the tank indicate a possible false cavity inside the tank.

6. Propane tanker—Look for the presence of discharge hoses. If none are present, find out how the propane is to be delivered. A lack of discharge hoses may indicate the lack of propane in the tank and that the truck has an alternative mission.

7. Concrete truck/mixer—Focus heavily on interview and on inspecting the four principal documents (insurance card, bill of lading, logbook, and driver's license) when inspecting this vehicle. Visually inspect the payload to ensure concrete is in the mixer. An issue indicator is if the mixer is not turning and vehicle accessories, water hose, chutes, etc. consistent with a concrete mixer are missing.

D.6.3.10 Checklist for the Inspection of Vehicles

Checklists for the inspection of vehicles are at the end of this appendix. Figure D-2 is a checklist for passenger vehicles, figure D-3 is a checklist for commercial trucks and trailers, and figure D-4 is a checklist for commercial cargo trucks. Use these checklists, combined with direction from regional and COCOM-specific guidelines and personal experience, to develop an installation-specific checklist.

PASSENGER VEHICLE INSPECTION CHECKLIST

General Observations

- [ ] **Look for anything in factory-built compartments**. When passenger vehicles are built, natural or factory-built compartments are present throughout the vehicle. These compartments can be excellent areas for hiding items.

- [ ] **Look for new or shiny bolts and/or screws**, indicating that something in that area has been altered, modified, or moved. Shiny bolts or screws indicate that the bolts or screws are new, where a scratched bolt or screw indicates removal and reinsertion.

- [ ] **Look for unusual scratches or other signs of tampering**. These may indicate that the area has possibly been modified and the repair or replacement was completed carelessly with some kind of tool. Notice if the sheet metal has any scratches and or small dents.

- [ ] **Check for unusually clean or dirty components in all areas.** Clean components in an otherwise dirty area indicate that these items are new or recently repaired or modified. ***The driver or owner of the vehicle should be able to explain why new or clean components are in the vehicle.*** Be especially aware of **clean or new wiring**, which can be directly connected to an explosive device.

- [ ] **Check for spools or remnants of electrical wire, tape, or similar items**. These materials are widely used in making vehicle bombs and may indicate that some type of device was constructed in the vehicle.

- [ ] **Look for new or broken welds**. A common technique in hiding items in deep concealment is to cover them with metal welded to the vehicle. Sometimes, existing welds are broken and rewelded to hide devices.

- [ ] **Look for unusual dirty or greasy fingerprints** on exterior surfaces. Residue from the hands is left in the general area where a person is working on a vehicle. When someone is working in an area, handprints and smudges can be left behind where road dirt has accumulated.

Inspecting the Exterior Front and Rear of Vehicle.

Have the driver hand you all required documents and ask the driver and any additional passengers to step out of the vehicle (See Paragraph D.6.4 for a list of required documents.). Have the driver open all closed vehicle compartments. Note: **Never open any vehicle compartments.**

- [ ] **Check the headlight/taillight wires.** The wiring for the lights can be used to connect an electrical source to a detonator.

- [ ] **Look for any fresh bodywork.** A close inspection of the body will reveal new bodywork. A typical way to hide devices is to create a false compartment in a section of the body. This false compartment is then covered with fiberglass, paint, undercoating, etc. (Be aware that in parts of Europe the undercoating is the consistency of grease.)

- [ ] **Feel or look at the back of the bumpers for a false compartment.** The bumper is typically made of steel plate and should not be overly thick.

- [ ] **Look in the area between the front grill and the radiator.** This area can be used to hold an explosive device. Electrical power from headlight wires is readily available.

Inspecting the Exterior Sides of the Vehicle

- [ ] **Look at the fenders for an unusually thick or wide area.** False compartments can be integrated into the fenders, especially in the areas above the tires. Look for fresh bodywork and undercoating.

- [ ] **Swing the doors.** Have the **driver initially open the doors**. Be aware of the feel of a normal door. If the door feels heavy when swung, then something may be hidden inside the door. The door cavity is an excellent place for hiding materials. Inspection holes are also found on the edge of the door by the latch. By removing the rubber plug, the inside of the door can be inspected. Rolling down a window is another way to search a vehicle's doors for concealed material.

Figure D-2. Passenger Vehicle Inspection Checklist (Sheet 1 of 4)

☐ **Tap on the bodywork along the side of the vehicle.** Using the knuckles of your hand, tap the vehicle. **(Do not damage the vehicle.)** Listen to the sounds. The length of the vehicle should sound hollow. If the sounds are inconsistent, there may be a change in density caused by material hidden inside the door(s).

☐ **Tap on the tires.** The tires should have a hollow or ringing sound. Any tire that sounds solid or does not sound hollow should be considered suspicious.

Inspecting the Engine Compartment

☐ **Take a moment to observe everything within view** for anything obviously unusual.

☐ **Check the battery box for anything unusual.** Some vehicles have large battery boxes that can be used for a smaller battery and have enough volume to act as a cavity for explosive materials.

☐ **Look for odd or additional wires running from the vehicle's battery.** Look for additional wires running from the headlight or the absence of a bulb in the headlight socket. Look for any unusual wiring. The battery is an excellent source of power for a detonator, and the vehicle wiring provides an easy way to connect the battery to the detonator.

☐ **Look for out-of place or unusually clean components, devices, and/or wiring and electrical tape.** Unusually clean areas, new tape, and/or new devices mean recent modifications or installations. The driver should be able to explain the clean areas and/or devices.

☐ **Inspect the firewall for any signs of modification or tampering.** Specifically look for signs of sheet metal work indicating the possibility of something hidden behind the firewall. Also look for new welds and new or shiny screws.

Inspecting the Vehicle's Interior

☐ **Take a few moments inside the vehicle to observe everything within view.** Pay particular attention to packages/devices (e.g., items with sound emitting from them, wires hanging from under the dash, loose carpeting or interior panels, and any iron or PVC pipes that may be in the interior of the vehicle).

☐ **Look closely at the dash for new, damaged, or scratched screws.** Behind the dash is a factory-built cavity. To obtain access, the dash must be removed. Removing and reinserting the screws can cause damage. A smuggler may decide to use new screws instead of reinstalling the old/ damaged ones.

☐ **Look for plugged vents on the dash.** The vents are a primary location for hiding moldable plastic explosives.

☐ **Look for bits of electrical tape, wire, stripped insulation, string, fine wire, fishing line, and/or any device such as a wrist watch that can be used as a time fuse, on the floor, dash, or seats.** All of these materials are commonly used in making explosive devices.

☐ **Check under floor mats for wires or switches.** Pressure switches that could be part of an explosive system are commonly used under floor mats and carpet.

☐ **Look at the glove box.** Be aware of the typical look and depth of a glove box for that type of vehicle. Be suspicious of any glove box that looks too small and/or shallow. A false compartment could be present.

☐ **Check all the seats.** Observe if there are any unusual bumps or bulges in the seats. If bumps and/or bulges are present, find out why they exist.

☐ **Check the roof liner.** Bulges, rips, frays, and/or repairs may indicate the possible concealment of an explosive device or explosive materials.

☐ **Look at the floor for anything that appears to be modified.** Remove floor mats if necessary. Typically, if anything is hidden in the floor, the floor will appear overly thick. Look for fresh welds or seams.

Figure D-2. Passenger Vehicle Inspection Checklist (Sheet 2 of 4)

☐ **Look for packages, containers, travel bags, and devices that seem out of place.** An explosive device will not necessarily be hidden in the vehicle. It may be hidden in a cardboard box on or under the seat.

☐ **Use your sense of smell inside the vehicle.** Be aware of the smell of caulking, glue, or any other unusual materials. Caulking, glue, and some other materials indicate that something has recently been modified or repaired. The driver should be able to identify any recent repairs.

Inspecting under the Vehicle.

Use a flashlight and mirror with a creeper (if possible) to carefully inspect under the vehicle. The development of search pits, such as those used in oil change and lubrication shops, is highly encouraged

☐ **Be sure all wire connections are properly complete.** The gas tank filler tube runs from the fill port to the actual gas tank with no bypasses, the exhaust pipe(s) runs from the exhaust manifold to the muffler(s), no loose wires (if so, determine why they are loose).

☐ **Look for anything taped or attached to the frame.** Nothing should be taped or wired to the frame. If so, have driver explain in detail.

☐ **Look for new welds on the frame.** Any new welds should be further investigated. The driver should know why the frame has new welds. New welds indicate that something has been added or modified.

☐ **Be suspicious of signs of recently installed hardware, such as a new muffler or fuel tank.** The driver should be able to explain any repairs or modifications to the vehicle.

☐ **Look for fresh undercoating or paint.** Fresh undercoating or paint may indicate that something is being concealed or hidden.

Inspecting the Cargo Area of Automobiles

☐ **Look for any new items in the trunk.** Notice if there is any new carpeting or mats. This indicates something has been changed recently.

☐ **Use your sense of smell when looking in the trunk.** Be aware of the smell of caulking, glue, or any other unusual materials. Caulking, glue, and some other materials indicate that something has recently been modified or repaired.

☐ **Move the trunk lid up and down.** Have the driver open the trunk for you. Notice if the trunk lid feels heavy, or if the trunk lid will not stay up. A heavy trunk lid indicates that something may be hidden in it.

☐ **Look for an unusually high or unusually thick trunk floor.** A false floor can be used to hide items below it.

☐ **Tap on the walls of the trunk to check for false compartments.** The sounds should be hollow. Any solid sounds indicate that something may be behind the wall and should be further investigated.

☐ **Look around the trunk area for unusual welds or seams.** Any of these could hide a loaded false compartment.

☐ **Check the area behind the rear seat.** This is an area where a false wall could be constructed.

☐ **Look at the spare tire area.** If the spare tire is not flush with the floor surface or does not look correctly installed, look for a false bottom to the tire well.

Inspecting the Cargo Area of Pickup Trucks

☐ **Check the weight of the tailgate by lifting it.** If it feels too heavy, something may be inside. Tailgates are generally hollow.

☐ **Look for fresh bodywork on the walls of the bed, the bed floor, and the tailgate.** A close inspection will reveal any new bodywork of fiberglass and paint. This work can be an attempt to hide a false compartment.

Figure D-2. Passenger Vehicle Inspection Checklist (Sheet 3 of 4)

☐ **Tap on the walls and bed of the cargo area to check for false compartments.** The sounds should be hollow. Any solid sounds indicate that something may be in the walls and should be further investigated.

☐ **Look around the bed area for unusual welds or seams.** These could be false compartments.

☐ **Look for signs of fresh caulking.** This could indicate that something has been recently modified.

☐ **Look in after-market stowage compartments.**

Inspecting Cargo Areas of Vans/SUVs.

The primary locations for hiding explosives in the cargo area are the walls of the bed, the tailgate, and in and under the floor.

☐ **Look for fresh bodywork.** Observe the walls, floors, and lift-back/doors. A close inspection will reveal any new bodywork of fiberglass and paint. This work can be an attempt to hide a false compartment.

☐ **Look around the floor area.** Check under the mat, if necessary, for unusual welds or seams. Any of these could be there to hide a loaded false compartment.

☐ **Look for signs of fresh caulking.** This indicates that something has been recently modified.

☐ **Check the ceiling for rips, bulges, or any signs of modifications for a false ceiling.**

☐ **Check the seat stow-and-go compartments if the vehicle is equipped with them.** Seat stowage compartments in the floor can be overlooked.

☐ **Check roof racks and or any other items attached to the roof of the vehicle.** Aftermarket devices can conceal explosives.

Figure D-2. Passenger Vehicle Inspection Checklist (Sheet 4 of 4)

SEMITRUCK VEHICLE INSPECTION CHECKLIST

General Observations

☐   **Look in factory- and aftermarket-built compartments.** When vehicles are built, natural or factory-built compartments are present throughout the vehicle. These compartments can be excellent areas for hiding items.

☐   **Look for new or shiny bolts and/or screws.** New or shiny bolts and/or screws indicate that something in that area has been altered, modified, or moved. The shiny bolts or screws indicate that the bolts or screws are new. A scratched bolt or screw indicates removal and reinsertion.

☐   **Look for unusual scratches or other signs of tampering.** Unusual scratches or other signs of tampering indicate that the area possibly has been modified and the repair or replacement was completed sloppily with some kind of tool. Look at sheet metal for tears, rips, and small dents.

☐   **Check for unusually clean or dirty components in all areas.** Clean components in an otherwise dirty area indicate that these items are new or recently repaired or modified. *__The driver or owner of the vehicle should be able to explain why new or clean components are in the vehicle.__*

☐   **Check for spools or remnants of electrical wire, tape, or similar items.** These materials are widely used in making vehicle bombs and may indicate that some type of device was constructed or installed in the vehicle.

☐   **Look for new or broken welds.** A common technique in hiding items in deep concealment is to cover them with metal welded to the vehicle. Sometimes existing welds are broken and re-welded to hide devices.

☐   **Look for oily or greasy fingerprints.** Working on most vehicles will get a person's hands oily or greasy. Residue from the hands is left in the area where a person has worked on a vehicle. When someone has worked in an area where road dirt has accumulated, handprints and smudges can be left behind.

Inspecting the Exterior Front and Rear of Vehicle.

Have the driver hand you all required documents (driver's license, registration, proof of insurance, and bill of lading). Have the driver and passengers step out of the vehicle. Have the driver open all closed vehicle compartments. **Note: The inspector should not open any vehicle compartments.**

☐   **Check the headlight and taillight wires.** The wiring for the lights can be used to connect an electrical source to a detonator.

☐   **Look around the fifth-wheel area.** The fifth wheel is the large plate behind the cab where the trailer is connected to the tractor. Look under this plate where it is attached to the chassis for signs of modifications. Be sure to check in any factory-built compartments.

☐   **Check for any compartments in the frame rails.** The frame rails are normally open. Compartments can be installed using welding techniques, so look for new welds there.

☐   **Look for any fresh bodywork.** A close inspection of the body will reveal any new bodywork. A typical way to hide devices is to create a false compartment in a section of the body. This false compartment is then covered with fiberglass, paint, undercoating, etc.

☐   **Check the vertical exhaust pipe for functioning and false compartments.** The exhaust pipe should be warm, or you should see the exhaust exit the pipe when the motor is running. Since the motor can run on only one exhaust pipe, the second pipe can be used as a hiding device.

☐   **Look in the area between the front grill and the radiator.** This area can be used to hold an explosive device. Electrical power from the headlight wires is readily available.

☐   **Feel or look at the back of the bumper for a false compartment.** The bumper is typically made of steel plate and should not be overly thick.

Figure D-3. Semitruck Vehicle Inspection Checklist (Sheet 1 of 4)

☐ **Look at the battery box for anything unusual.** Ensure that all batteries appear to be connected properly. Empty battery shells can be used as false compartments.

☐ **Look in storage compartments for anything unusual.** Typical items include tools, rags, motor oil, chains, chocks, etc. These compartments are usually dirty.

☐ **Look at the fenders for any unusually thick or wide area.** False compartments can be integrated into the fenders, especially in the areas above the tires. As before, look for fresh bodywork and undercoating.

☐ **Tap on the tires.** The tires should have a hollow or ringing sound. Any tire that sounds solid or does not sound hollow should be considered suspicious.

Inspecting Fuel Tanks.

Fuel tanks are a primary area for hiding explosives.

☐ **Look for unusual welds under the holding straps.** The holding straps offer cover for new welds. Look on top of the tank where the straps are not in contact with the tank. You should be able to see any welds in this uncovered area.

☐ **Tap on the tanks and listen for inconsistent sounds.** If there is something in the tanks other than fuel, you will hear different sounds on different areas of the tanks.

☐ **Look for missing connections on the tanks.** If something other than fuel is in the tank, the smuggler may not bother hooking the fuel lines to the tank.

☐ **Feel the tanks with your hand for temperature inconsistencies.** Since the height of the fuel will be level, the temperature should be the same along the length of the tank, whether you are touching the filled or unfilled volume.

Inspecting the Engine Compartment.

Take a moment to observe everything within view for anything obvious or unusual.

☐ **Look for odd or additional wires running from the vehicle's battery.** Look for additional wires running from the headlight or the absence of a bulb in the headlight socket. Look for any unusual wiring. The battery is an excellent source of power for a detonator, and the vehicle wiring provides an easy way to connect the battery to the detonator.

☐ **Look for out-of-place or unusually clean components, devices, and/or wiring and electrical tape.** Unusually clean areas, new tape, and/or new devices mean recent modifications or installations. The driver should be able to explain clean areas and/or devices.

☐ **Check the windshield washer container for a false compartment or signs of tampering.** This container could contain a significant amount of explosive material. Ensure that the windshield washer hose is properly hooked up to the system. Open the container and smell the contents. This container could also be a false fuel tank. If it is, the fuel tanks may contain something other than fuel.

☐ **Check under larger components.** Look at the air cleaner and fan blade shrouds for unusual containers. If the container is legitimate, the driver should be able to explain its purpose. The container could carry explosive materials or fuel. If a vehicle has to move only a short distance, these containers can be used as a short-range fuel tank. Open the containers if possible and smell the contents. If it is fuel, then the fuel tanks on the vehicle are probably being used for foreign materials. Look for wires going into the container. These wires could be connected to a detonator.

☐ **Inspect the firewall for signs of modification or tampering.** Especially look for signs of sheet metal work, indicating the possibility of something hidden behind the firewall.

Figure D-3. Semitruck Vehicle Inspection Checklist (Sheet 2 of 4)

Done below:

Inspecting the Inside of the Passenger Compartment.

Take a moment to observe everything within view. Pay close attention to packages/devices (e.g., alarm clock, iron or PVC pipe) that look out of place.

- [ ] **Look closely at the dash for new, damaged, or scratched screws.** Behind the dash is a factory-built cavity; to get access, the dash has to be removed. Removing and reinserting the screws can cause damage. A terrorist may decide to use new screws instead of reinstalling the old/damaged ones.
- [ ] **Look for plugged vents on the dash.** The vents are a primary location for hiding plastic explosives.
- [ ] **Look for bits of electrical tape, wire, stripped insulation, string, fine wire, fishing line, and/or a time fuse on the floor, dash, or seats.** All of these materials are commonly found in explosive devices.
- [ ] **Check under the floor mats.** Check for wires or switches that could be part of an explosive system.
- [ ] **Look at the glove box.** Be aware of the typical look and depth of a glove box. Be suspicious of any glove box that looks small and/or shallow. A false compartment could be present.
- [ ] **Check the seats.** Especially check the passenger seat for unusual bumps or bulges. If bumps and/or bulges are present, find out why they exist.
- [ ] **Check the roof liner.** Check for bulges, rips, and/or repairs indicating possible concealment of an explosive device or explosive materials.
- [ ] **Swing the doors.** Have the driver initially open the doors. Be aware of the feel of a normal door. If the door feels heavy when swung, something may be hidden inside. The door cavity is an excellent place for hiding materials.
- [ ] **Look at the floor for anything that appears to be modified.** Remove floor mats if necessary. Typically, if anything is hidden in the floor, the floor will appear overly thick. Look for fresh welds or seams.
- [ ] **Be suspicious if the interior is unusually clean.** Most trucks are used daily and are not kept in clean condition.
- [ ] **Look for the presence of personal items.** Know the purpose of the incoming vehicle. If the truck has been on a long haul, then the driver should have a suitcase and/or similar items present. If there is a lack of these types of items, the driver should be able to explain the reason.
- [ ] **Look for packages, containers, travel bags, and devices that seem out of place.** An explosive device will not necessarily be hidden in the vehicle. A cardboard box on the seat, in the sleeper, or under the seat may be the device you are looking for. Also, truck travelers typically carry soft luggage due to the conditions in the sleeper. Therefore, be suspicious of a hard suitcase.
- [ ] **Use your sense of smell in the sleeper area.** Be aware of the smell of caulking, glue, or any other unusual materials indicating that something has recently been modified or repaired. The driver should be able to identify any recent repairs. Also, the presence of a large number of air fresheners may be an indicator that the driver is attempting to cover a strong smell — perhaps from an explosive device.
- [ ] **Look under the mattress for any unusual items.** Personal items are typically stored in this area. Be suspicious of any non-personal items.
- [ ] **Check the walls and ceiling of the sleeper.** These should not be overly thick.

Figure D-3. Semitruck Vehicle Inspection Checklist (Sheet 3 of 4)

Inspecting Under the Vehicle.

Always use a flashlight and mirror with a creeper (if possible) to carefully inspect under the vehicle.

☐ **Be sure all connections are properly made.** Check that the gas tank filler tube runs from the fill port to the tank and the exhaust pipe runs from the manifold to the muffler.

☐ **Look at the frame for signs of a false compartment.** The frame rails are typically made of C-channel and do not contain any built-in compartments.

☐ **Look for anything taped or attached to the frame.**

☐ **Be suspicious of signs of recently installed hardware.** The driver should be able to explain any repairs or modifications to the vehicle.

☐ **Look for fresh undercoating or paint.** This may indicate something is hidden.

☐ **Look for spare or extra tanks on the vehicle that have no obvious uses.** A smuggler may add tanks and supporting lines and connectors to the vehicle for the purpose of hiding explosives.

Inspections of Trailer Undercarriages

☐ **Look at the fifth-wheel area for false or loaded compartments.** The fifth-wheel area contains a steel plate attached to the frame of the trailer. A factory-built cavity is between this plate and the trailer frame. This area should be free of any material. Check to see if additional plates have been welded in this area to create compartments above the steel plate.

☐ **Check the floor for unusual thickness or evidence of modifications.** Evidence includes new floorboards, damaged floorboards, new or shiny screws, and an unusually high floor level.

☐ **Look at the under-structure of the trailer.** The frame should not contain any compartments or boxed-in areas. Note: Outside the continental United States, trailer frames have a large box beam as the support. This box beam is hollow and can be used to smuggle material.

☐ **Look at the thickness of the floor from below.** Know where the bottom of the floor should be with respect to the frame rails.

☐ **Be suspicious of new tires on the trailer.** Most operators do not buy a complete set of new tires. Typically, trailer tires are heavily worn or are retreads.

☐ **Look for tires obviously smaller than the adjoining tires.** The tire presents a good area for hiding material. However, a spinning tire presents a hot, hostile environment for explosives or explosive devices. A smaller tire will not spin and, therefore, will be a suitable environment for smuggling material. The smaller tire will almost always be on the inside of the axle.

☐ **Look for new or strange air tanks.** An air tank is necessary for braking the vehicle and is typically near the rear axle of the trailer. This tank should be pressurized. Use the bleeder valve to check for air pressure. If there is no air pressure, the tank may be being used for other purposes. Be suspicious of any other tanks on the vehicle. The driver should be able to explain the purpose of the other tank(s).

Figure D-3. Semitruck Vehicle Inspection Checklist (Sheet 4 of 4)

CARGO VEHICLE INSPECTION CHECKLIST

Inspection of Semi-Van Trailer/Straight Box Truck/Step Van.

The typical use for these box-like closed cargo vessels is transporting all types of merchandise and commodities.

Trailer (Outside)

☐ **Look at the company insignia and other items applied to or painted on the surface of the vehicle.** All items should be professionally applied. Be suspicious if the company logo and other supporting information are painted or applied poorly.

☐ **Look at the walls for evidence of modifications.** The walls of the trailer are typically constructed using plywood panels held together by rivets and/or screws. Look for possible false walls in the trailer. Items to look for include inconsistent spacing of wall panels; missing rivets or screws; and areas with shiny, damaged, or new-looking screws or rivets. In addition, look for misaligned or overlapped seams at the exterior roof/wall intersection and new plywood or other materials in the walls.

☐ **Look for false walls in the front of the cargo area.** A false wall can be determined by comparing the exterior length of the trailer to the length of the cargo area. The lengths can be measured accurately with a measuring tape or less accurately by pacing the distances.

☐ **Check the ribs in the walls**. Count the number of ribs on the exterior wall of the trailer and compare that number to the number of ribs on the wall in the cargo area. An inconsistent count indicates a false wall in the front of the cargo section.

☐ **Check for false roofs on the top of the trailer.** Sometimes, the top of the trailer will be loaded with explosives. Access to this area is by hinging the roof of the trailer to the wall.

☐ **Look for hinges or signs of tampering.** Check where the roof and side walls meet.

☐ **Check the floor surface for signs of modifications.** Be suspicious of new or modified floor planks that may cover a hiding area below. Be aware of the normal height of the floor and look for anything unusual.

☐ **Tap on the walls and roof to discover false cavities.** Listen for hollow sounds since the roof and walls are typically made of single-layered materials and should not sound hollow.

Inspection of Cargo in the Trailer

☐ **Shine a light under the cargo pallets.** You should be able to see all the way to the front wall of the cargo area. If you can't, then an unusual item or items are present, or the floor may have been modified. Determine what the cause is.

☐ **Cargo typically consists of a single uniformly boxed commodity.** Therefore, when looking over the top of the cargo, the top surface of the load should be even. If there are dips or packages above the normal cargo level, determine the reason for the non-uniformity. Something could be hidden in this area.

☐ **Notice where the cargo boxes were purchased.** Companies usually purchase large quantities of boxes with their logos on the side for economic reasons. Be suspicious if the cargo is boxed in U-Haul or similarly named boxes since these boxes are expensive and shipping companies tend to avoid buying them.

☐ **Look at the way the cargo is stacked.** It should be neatly stacked on pallets so it can be handled easily. A sloppily stacked cargo or a cargo of many different items could mean lack of interest in the safety of the cargo and that it is there strictly for appearance and/or diversionary reasons.

☐ **Look over the top surface of the cargo.** Look for any signs that someone has walked on top of the cargo.

☐ **Notice the shipping labels on packages hauled by the popular shippers.** Most, if not all, of the packages and boxes in the cargo area should have the shipper's name clearly on them.

Figure D-4. Cargo Vehicle Inspection Checklist

D.6.4 Visitor Passes

1. Authorized personnel or sponsored visitors without an approved identification card desiring to enter a Navy installation will be required to obtain a visitor's pass. Sponsored visitors must be verified with the sponsor.

2. Authorized or sponsored visitors desiring to enter a Navy installation that require a visitor's pass shall:

   a. Show a valid driver's license and vehicle registration certificate/rental vehicle agreement.

   b. Fill out a log that include names, home address, destination, make and model of vehicle, license plate number, sponsor, time of arrival, and estimated time of departure.

3. A debarment list shall be prepared by the installation SJA, provided to the security department and shall be checked prior to the issuance of a visitor pass.

4. All visitor's passes shall be strictly controlled. Unused passes will be securely stored to prevent theft.

D.6.5 Unauthorized Pedestrian or Vehicle

Whenever possible, use barriers to channel, slow, and physically stop the vehicle until it can be cleared to enter. Each ECP must contain the specific, preplanned response that the sentry will execute. The decision to authorize the use of deadly force will be influenced by the following:

1. The justifications for deadly force as detailed in the rules for the use of deadly force.

2. The conspicuousness of the warning to stop.

3. The potential targets protected by the ECP, such as sensitivity, classification, and operational importance of the area, as well as the nature of the potential targets inside (e.g., high-ranking personnel and large-occupancy buildings).

4. Latest threat warning and current FPCON.

5. The number of, and means by which the suspect negotiated, the obstacles or barriers protecting the ECP.

6. Any indications that the vehicle poses a specific threat capability (e.g., visible explosives or symbols, visible weapons, type of vehicle, and HAZMAT symbols).

7. Indications of intent, such as evident duress or nervousness of suspect(s), threatening statements, clothing (e.g., hoods, masks, body armor, explosive vests).

No sentry (unless operating under the ROE/RUF that state otherwise) shall engage a driver or vehicle with weapons fire unless the sentry determines that conditions of extreme necessity exist and he is justified in using deadly force as described in SECNAVINST 5500.29 (series), Use of Deadly Force and the Carrying of Firearms by Personnel of the Department of the Navy in Conjunction with Law Enforcement, Security Duties, and Personal Protection.

Note

When the decision to use deadly force is made, direct fire at tires, engine, windshield, and driver. Depending on the circumstances, these decisions are not made in any particular order.

When the RRP do not authorize deadly force, the entry control guard should:

1. Immediately notify a supervisor and provide the description of the vehicle and the direction in which the vehicle was proceeding.

2. Immediately secure the ECP.

D.6.6 Protests and Rallies

The SO should develop preplanned responses for protests and rallies at ECPs, to include all required installation response requirements. If protestors approach an ECP, watch standers should execute the following procedures:

1. Upon initial gathering or massing of personnel, notify COG/patrol supervisor or ECP supervisor.

2. Prepare to secure ECP by closing gates or activating barrier systems.

3. Maintain the flow of friendly personnel through the ECP.

4. Do not take actions that would escalate the situation.

5. Request reinforcements as required.

6. Verbally warn protestors to remain away from the post as directed by supervisors.

7. If protestors physically attack a watch stander or attempt to forcefully gain access to the protected area:

    a. Prevent access by using authorized force.

    b. Apprehend aggressors and remove from immediate area.

    c. Secure ECP as directed to prevent unauthorized access.

    d. Assist civilian authorities as required.

    e. Maintain positive access control at all times.

8. An indicator that a protest or rally is imminent at a particular ECP is the unexpected arrival of news organizations as they often have advanced notice of groups' intentions. Immediately notify the COG/patrol supervisor or ECP supervisor if this occurs.

D.6.7 Media Requests

Representatives of the media shall not be allowed access to the protected area without proper authorization. Additionally, watch standers shall:

1. Notify supervisor when media representatives identify themselves, attempt to gain access to the protected area, or are visibly present near the ECP (displaying cameras, lighting or sound equipment, or marked vehicles).

2. Refer all media queries to proper military authority, normally the PAO/command spokesperson, CDO, executive officer, or commanding officer.

D.6.8 Presence of Improvised Explosive Device

If an IED is discovered at an ECP, watch standers will take the following actions:

1. Communicate presence of suspect item to watch team present at location.

WARNING

Do not use an electronic device such as a radio to communicate the presence of a suspect item as this may detonate the device (radio signals could detonate explosives).

2. Alert COG/patrol supervisor or ECP supervisor using a preplanned procedure that does not involve using a radio.

Note

Do not use an electronic device such as a radio to communicate the presence of a suspect item as this may detonate the device.

3. Verbally command the suspect to move away from the suspected item. If suspect is uncooperative, physically take control of suspect and move away from the suspected item.

4. If the suspect is wearing the IED, attempt to control the suspect with verbal commands and have the suspect move to a safe position that allows for standoff distances for bystanders and watch standers. Do not attempt to physically separate the suspect from the IED. Watch standers must be aware of the RUF in these situations.

5. Once the suspect is separated from the IED, restrain and search suspect.

6. Secure ECP by closing gates and activating barriers.

7. Direct all unarmed personnel to stand clear at specified distance.

8. All armed watch standers fall back to a safe distance within the effective range of their small arms and wait for EOD.

D.6.9 Unattended Package at or Near the Entry Control Point

Watch standers shall take the following actions when an unattended package is noticed at or near the ECP:

1. Communicate presence of suspect item to watch team.

2. Alert COG/patrol supervisor or ECP supervisor using a preplanned procedure that does not involve using a radio (radio signal could detonate explosive).

3. Secure ECP by closing gates and activating barriers.

4. Direct all personnel to stand clear of suspected IEDs or explosive material at specified distances.

D.6.10 Surveillance Detection

When there is suspicion or detection that the ECP or other asset is being observed, NSF should not engage suspected surveillance attempts. NSF should attempt to collect as much information as possible IAW HHQ guidance. NSF should contact NCIS and/or local law enforcement to engage IAW established MOU/MOA. Watch standers shall:

1. Note suspect's physical characteristics, vehicles and license plate numbers (if present), methods of surveillance (cameras, video, binoculars), and exact location.

2. Notify COG/patrol supervisor or ECP supervisor via means other than a radio so as not to alert observers who might be monitoring radio frequencies.

3. Maintain visual contact and report any changes in location or suspect descriptions until arrival of responding personnel.

D.6.11 Medical Emergency

If there is a medical emergency at or near the ECP, watch standers shall:

1. Determine whether there is a legitimate medical emergency and not a pretense for terrorists to drive an ambulance onto an installation.

2. Positively identify the emergency responders.

3. Communicate situation to COG/patrol supervisor or ECP supervisor.

4. Determine whether victim requires first aid or EMS.

5. Maintain positive control of ECP at all times. Secure post only if necessary to provide first aid to victim.

6. Direct responders to victim.

INTENTIONALLY BLANK

APPENDIX E

Traffic Enforcement

E.1 GENERAL

The purpose of traffic LE is to reduce traffic accidents through preventive patrol and active enforcement. To effectively carry out this function, all officers must be familiar with applicable federal, as well as assimilated state, statutes that apply to traffic LE on their installation. The SO exercises overall responsibility for directing, regulating, and controlling traffic and enforcing laws pertaining to traffic control aboard the installation with exclusive jurisdiction. An MOU should be agreed upon with local LE in areas of concurrent or proprietary jurisdiction.

NSF will provide traffic enforcement IAW AR 190-5/OPNAV 11200.5D/AFI 31-218(1)/MCO 5110.1D/DLAR 5720.1, Motor Vehicle Traffic Supervision, and DODD 5525.4, Enforcement of State Traffic Laws on DOD Installations; or if outside the continental United States (OCONUS), to the extent military authority is empowered to regulate traffic. NSF should have complete working knowledge of all traffic codes applicable to the installation under the applicable SOFA, including:

1. The rules of the road, parking violations, towing instructions, safety equipment requirements, and other key provisions.

2. Applicable portions of the Uniform Vehicle Code and Model Traffic Ordinance published by the National Committee on Uniform Traffic Laws and Ordinances (23 Code of Federal Regulations (CFR) 1204).

E.2 EMERGENCY RESPONSE

Emergency response to serious incidents or threats should be made only in emergency vehicles. Emergency vehicle response shall be authorized only in those situations identified in Chapter 3 of this instruction. Emergency vehicles shall be properly equipped for emergency response with all lights, sirens, radio, and PA systems. NSF shall use extreme caution and emergency vehicle equipment when conducting an emergency response. The deactivation of emergency equipment prior to arrival at the scene may be appropriate in certain tactical situations.

E.3 TRAFFIC ENFORCEMENT TECHNIQUES

Traffic law enforcement should motivate drivers to operate vehicles safely within traffic laws and maintain an effective and efficient flow of traffic. Effective enforcement should emphasize voluntary compliance by drivers and can be achieved by the following actions:

1. Visible traffic patrol.

2. Stationary observation.

3. The tendency of motorists to knowingly violate traffic laws is deterred by open and visible patrol. However, when there is an unusual or continuing enforcement problem at a particular location, officers may park in a conspicuous location and observe traffic.

4. Only NSF that have been specially trained and certified IAW local jurisdiction requirements are allowed to use radar and speed-measuring devices.

a. The equipment shall be calibrated and maintained as specified by the manufacturer and applicable state requirements. Records of the maintenance, calibration, and repair of radar units shall be retained for two years.

b. An external test for accuracy of each unit's operation must be conducted before the unit is placed in operation. If a test is conducted after each violation, it should be recorded on the file copy of the traffic ticket.

c. Operators must demonstrate proficiency with radar equipment. Training programs must take into consideration local, legal, and certification requirements. In addition, hands-on training in the field and a reasonable period of practice shall be conducted under the guidance of an FTO before an operator is allowed to issue a traffic ticket.

E.4 TRAFFIC ACCIDENT RESPONSE

The officer's response to an accident scene will be determined by the magnitude of the accident as reported. Officers responding to the scene of any accident shall drive in a safe manner with due regard for persons and property. Emergency lights and siren shall be used when responding to accidents with known or probable injuries. Upon arrival, the officer will determine whether additional assistance is required at the scene.

E.5 TRAFFIC DIRECTION AND CONTROL

E.5.1 Manual Direction

Officers will manually direct traffic under the following circumstances:

1. During periods of traffic or pedestrian congestion where traffic control signals are malfunctioning

2. During special events (notification should be given in advance of any planned special event)

3. Before and after school at crossing zones that do not have guards assigned.

E.5.2 Use of Hand and Arm Signals

1. Assume the basic stance.

   a. Stand with feet about shoulder width apart so weight is evenly distributed. To avoid passing out, do not lock the knees or tense the body.

   b. Let hands and arms hang naturally.

   c. Hold head and body in a disciplined manner.

2. Direct traffic from the right to proceed straight ahead.

   a. Look toward the traffic.

   b. Extend the right arm up and out to the right side parallel to the ground. The hand is straight, palm up, and fingers extended and together.

   c. With the elbow as a fixed axis, rotate the right forearm across the front of the body so the hand stops just below the chin with the palm facing down. To avoid confusion to motorists, do not move any other parts of the body.

   d. Complete the signal by dropping the arm smartly and resuming the basic stance.

3. Direct the left traffic straight ahead. This step is the same as step 2, but substitute left for right as in 2b.

4. Direct the right traffic in a right turn.

   a. Look toward the traffic.

   b. Extend the right arm toward the traffic, parallel to the ground. Point the first two fingers, palm facing forward, at the vehicle waiting to turn.

   c. In a sweeping motion, bring the hand around to point in the direction the traffic wishes to go.

   d. Return to the basic stance.

5. Direct the left traffic to turn right.

   a. Look toward the traffic.

   b. Extend the left arm toward the traffic, parallel to the ground. Point the first two fingers, palm facing forward, at the vehicle waiting to turn.

   c. Without turning the shoulders or body, sweep left arm 8–10 inches to the rear (to the rear, i.e., the driver's right).

   d. Return to the basic stance.

6. Direct the right traffic to turn left.

   a. Halt traffic on the left, allowing the driver time to react and bring the vehicle to a stop.

      (1) Look to the left.

      (2) Thrust the left hand toward the traffic, with the hand flat, palm facing traffic and above the headgear.

   b. Holding the stop signal, extend the right arm toward the traffic parallel to the ground. Point the first two fingers of the right hand at the vehicle waiting to turn.

   c. Without turning the shoulders or body, sweep the right arm 8–10 inches to the rear to signal the traffic to turn.

   d. Drop the right arm to the side and, after the vehicle has cleared the intersection, look back toward the waiting traffic on the left.

   e. Change the left-hand stop signal to a come-through signal by rotating the left palm in and the left forearm across the front of the body until the hand is just below the chin. Finally, drop the arm and resume the basic stance.

7. Direct the left traffic to turn left.

   a. Halt traffic on the right the same as one would on the left.

   b. Holding the stop signal, extend the left arm toward the traffic, parallel to the ground. Point the first two fingers of the left hand at the vehicle waiting to turn.

   c. In a sweeping motion, bring the hand around, pointing in the direction the traffic wishes to go.

    d. Drop the left arm to the side and, after the vehicle has cleared the intersection, look back toward the waiting traffic on the right.

    e. Change the right-hand stop signal to a come-through signal by rotating the right palm in and the forearm across the front of the body until the hand is just below the chin. Then drop the arm and resume the basic stance.

8. Stop traffic from the front.

    a. Raise the left arm up and out to the front. Keep the fingers together and extended with palm facing out so the entire hand is clearly visible to oncoming traffic. The elbow should be bent at eye level so the hand is above the headgear.

    b. When the traffic is halted, drop the arm and return to the basic stance.

9. Stop traffic from the rear.

    a. Slightly bend the left knee, twist to the right, and turn the head and eyes to the rear. Do not move the feet.

    b. Raise the right arm up and out to the rear, fingers extended and together, palm facing traffic. Keep the elbow bent and at eye level.

    c. When traffic is stopped, complete the signal by dropping the arm and returning to the basic stance.

10. Change the flow of traffic.

    a. Look to the right, put up a stop signal with the right arm, and hold that signal.

    b. Do the same with the left arm.

    c. Turn the body 90 degrees to face the traffic on the right or the left, facing the traffic just stopped.

    d. If to the left, convert the left-arm stop signal to a come-through signal by rotating the palm 180 degrees to face inward. With the elbow fixed, rotate the forearm across in front of the body until the hand is just under the chin, palm facing down. Complete by dropping the arm.

    e. If to the right, do the same as d. above but use the right arm.

    f. Return to the basic stance.

E.5.3 Fire and Emergency Scenes

Officers directing traffic at fire and emergency scenes shall ensure that private vehicles are well clear of the emergency scene and are not obstructing emergency vehicles or other traffic.

E.5.4 Road Hazards

Officers should report any road hazards to dispatch. The following are considered road hazards:

1. Damaged or malfunctioning traffic control devices

2. Defective roadway lighting

3. Visually obscured intersections

4. Roadway defects

5. Lack of, damaged, or missing roadway signs or safety devices.

E.6 ACCIDENT INVESTIGATIONS

The investigation of traffic accidents is necessary not only to determine traffic law violations, but also to obtain engineering data, protect the rights of the individuals involved, and assist in traffic education. To ensure proper and complete investigation of accidents, the following procedures will be used.

1. Upon arrival at an accident scene, officers will:

    a. Park the patrol vehicle so as to protect the scene and allow movement of traffic if feasible.

    b. Wear the issued reflective vest when working accident scenes.

    c. Administer first aid and notify dispatch when rescue and/or wrecker service is needed.

    d. Request another patrol if needed for assistance.

    e. Set flares or reflective triangles as needed.

WARNING

Warning flares should never be used if any type of fuel is present in the area.

2. On-scene procedures.

    a. When serious bodily injury, death, or extenuating circumstances exist, the installation photographer is to be called. In this case, vehicles should not be moved unless absolutely necessary to preserve life or prevent further collisions.

    b. Obtain driver's license, vehicle and installation registration, and proof of insurance from all drivers involved in the accident.

    c. Question and obtain names and addresses from any witnesses. When it is necessary for a witness to leave the scene before the investigation is complete, obtain all necessary information as quickly as possible and allow the witness to depart.

    d. Investigate and determine the cause of the accident. Note the position of all vehicles involved and take measurements whenever possible.

    e. After the preliminary investigation is completed, clear the roadway quickly and refrain from blocking any portion of the roadway while completing paperwork.

    f. After the roadway is clear, the investigating officer should turn the unit's blue lights off as quickly as possible if this can be done without creating a hazard. This action will usually allow traffic to flow faster by attracting less attention.

    g. When there are traffic violations, issue the citation as appropriate and allow the drivers to leave.

    h. If the driver and passengers of any vehicle involved were transported from the scene because of injuries, the officer will follow up, obtaining all the information necessary to complete the investigation and report. Where injuries are minor and all of the needed information has been obtained at the scene, it is not normally necessary for the officer to conduct a follow-up investigation.

E.7 ACCIDENTS INVOLVING SERIOUS INJURY OR FATALITY

1. If an accident investigator is responding to a serious injury or fatality accident:

    a. The NSF receiving the initial call is to park outside the accident scene. Dispatch shall immediately notify the on-call NCIS special agent in charge whenever there is a fatality involved.

    b. The first NSF on the scene shall advise dispatch of the emergency equipment needed and begin administering the appropriate first aid to the injured.

    c. When victims are transported from the scene, the patrol supervisor or the accident investigator shall direct emergency medical technicians or ambulances into and out of the area without disturbing the scene if at all possible.

    d. After survivors have been removed from the scene and the accident investigation team has been requested to handle the case, the accident scene shall be protected by diverting traffic from the area. Under no circumstances will wreckers or spectators be allowed to enter the accident scene unless authorized by the senior accident investigator.

    e. The accident investigator shall relay all pertinent information to the initial officer. The accident investigation team shall be in charge of the accident scene and will make the determination of what course of action to take as to when to remove vehicles; make photographs; order blood, breath, or urine samples; etc.

2. If a trained accident investigator is unavailable or unable to respond, the following steps shall be taken.

    a. The initial NSF responding to the call will arrive on the scene and park outside the scene area.

    b. Upon determination that a fatality exists, the initial NSF shall advise the dispatch and call for assistance.

    c. Call for the installation photographer and other emergency assistance as needed. The accident scene shall be protected as a crime scene. Other traffic shall be diverted or directed around the scene.

    d. Any suspects at the scene shall be detained by the initial NSF, either at the scene or, if injured, at the medical facility for later investigation.

    e. No items such as vehicle parts, body limbs, or deceased persons should be disturbed or removed from the scene if at all possible.

    f. No wreckers or spectators will be allowed to enter the scene until authorized.

    g. If a fatality exists, as determined by a certified or appropriate authority, the victim should not be removed from the scene. Ensure NCIS is notified for all fatalities. However, if a victim must be removed, the responsible NSF shall document the position of the victim before removal.

    h. The appointed investigating NSF shall:

        (1) Assume command of the accident scene. All pertinent information shall be relayed to this investigator.

(2) If different from the initial responding NSF, the investigating NSF should be called by the supervisor on the scene. The initial responding NSF shall make a supplemental report outlining his/her activities to be included in the investigative report.

(3) Complete all investigative reports and be responsible for conducting and concluding the investigation, including the initiation of any criminal charges that may be forthcoming and ordering blood, breath, or urine samples for testing.

(4) Be responsible for clearing the accident scene, with the exception of biohazardous material, and impounding vehicles if required for further investigation. The wrecker service removing the vehicle(s) will clear the roadway at the accident scene.

(5) Not release any accident details and shall direct all inquiries for news releases to the PAO.

(6) Submit all following reports concerning fatalities.

    (a) Vehicle accident report

    (b) IR, outlining in detail the complete investigation

    (c) Vehicle impound report

    (d) Copies of witnesses' statements

    (e) Blood alcohol test on victim and suspect (if applicable)

    (f) Copies of citations issued

    (g) Reports by medical examiner or coroner

    (h) Photographs (to be attached later).

E.8 HIT AND RUN (LEAVING THE SCENE OF AN ACCIDENT)

1. First NSF to arrive on the scene:

    a. Administers first aid and advises the dispatch when additional patrols and/or rescue/wrecker service is needed.

    b. Obtains information and advises dispatch to issue a be-on-the-lookout (BOLO) for the suspect's vehicle.

    c. If the hit-and-run accident involves a fatality, advises dispatch for the CDO to initiate notification procedures.

2. Conduct operations and investigations as listed in E.7 and E.8 of this appendix as they apply and:

    a. Collect evidence that would aid in identifying the suspect's car.

    b. Secure and provide any evidence collected at the scene into the evidence custody system.

    c. Provide additional/emerging information for BOLO broadcast.

    d. Complete a vehicle accident report and mark clearly on that report that the accident is a hit-and-run.

E.9 WRECKER SERVICE

1. Requests for wrecker service should go through dispatch.

2. The NSF patrol, not the wrecker driver, is responsible for filling out the vehicle inventory form.

   a. An inventory form shall be completed on every vehicle towed.

   b. Two copies of the inventory sheet will be needed—one to the wrecker service and one to be turned over to the records office.

3. NSF shall remain at the scene until the wrecker has towed the car away.

4. Accident victims may use the wrecker service of their choice when:

   a. The victim or driver is not under apprehension.

   b. The vehicle involved is not causing an immediate traffic hazard and will be moved within a reasonable time.

5. When a wrecker is required and the driver/victim is able to communicate, the patrol should determine whether the driver wants a wrecker for his/her vehicle. The driver should be made aware the wrecker service cannot be canceled once the wrecker is dispatched.

E.10 LOW-/HIGH-RISK TRAFFIC STOPS

1. NSF should conduct low-risk traffic stops in the event of traffic violations or defective equipment or to investigate suspicious activity. Local procedures shall be established stressing safety, timeliness, and courtesy.

2. NSF shall conduct high-risk traffic stops when there is a reasonable belief that the driver or occupants have committed a serious criminal offense and there is an imminent threat to the safety of the NSF personnel or the public.

3. NSF should avoid high-speed pursuits aboard the installation due to potential danger to NSF, bystanders, and suspects.

4. Moving roadblocks shall not be used. On those occasions when stationary roadblocks are necessary, they will be established only IAW with RRP and this NTTP.

   a. Roadblocks at entry and exit points of an installation should be used only when the fleeing vehicle could cause serious harm to the public.

   b. Ramming a suspect vehicle by NSF vehicles is prohibited.

   c. The use of tire-flattening devices is authorized.

   d. SOs will establish local written policy regarding the use of roadblocks on the installation.

5. Vehicle checks/suspicious vehicle—felony stops

   a. Felony stops are extremely dangerous and carry a greater possibility of risk for NSF personnel. The vehicle may have been identified as stolen and/or the individuals in the vehicle may be suspected of a serious offense. When making felony stops of suspicious vehicles, officers are to adhere to the following procedures.

(1) NSF shall notify dispatch when they are following a vehicle they wish to investigate. The patrol should give dispatch the license number, a description of the vehicle, the direction of travel, approximate location, and the number of occupants, including gender and race.

(2) A primary response unit—assisted by one or, if needed, more secondary (backup) units—will normally conduct felony stops.

(3) If the vehicle stops unexpectedly and the occupants attempt to exit the vehicle; the NSF is to order them to remain in their vehicle. If a backup unit is en route, the NSF will remain with his/her patrol unit until the arrival of the backup unit.

(4) Backup units are to stop to the right of the first patrol unit. The backup unit's headlights should be left on and the emergency flashers and overhead emergency lights left on.

(5) Stop vehicles by using lights and/or siren or public address system. The patrol vehicle should be approximately 30 feet behind and 2–3 feet to the left of the suspect vehicle.

(6) When possible, patrols should stop suspicious vehicles in well-lighted areas. Additionally, they should avoid stopping the vehicle at an intersection or in a heavily congested area.

(7) An inquiry on the license number of the suspicious vehicle should be made and the results reported immediately back to the NSF patrol when available.

(8) Maximize cover/concealment by opening doors and ensuring access to the radio and PA system. If the primary NSF vehicle is a two-man unit, the driver shall operate the PA system and the second officer will operate the radio, maintaining radio contact with dispatch.

(9) The initiating patrol shall establish contact with the occupants using the PA system. Use a command voice if a PA is not installed. The patrol shall identify themselves and the reason for the stop. If the stop is occurring at night, have the suspect vehicle driver turn on the dome light.

(10) Do not begin to remove occupants until backup unit(s) is (are) on the scene and in position.

(11) The initiating patrol communicates with the suspect(s). All responding patrols shall have their weapon out and finger straight along the trigger housing (unless hostile situation is evident) while covering the suspect vehicle.

(12) The initiating patrol shall first have the driver take the keys from the ignition with the right hand and hold the keys out of the window in their right hand while exposing both hands and then instruct the driver to get out of the vehicle on the driver's side, raising his or her arms to shoulder level and turning around 360 degrees. The driver will be instructed when to drop the keys so it drops in an area where the keys will not be lost or accidently dropped in a storm water drain.

(13) The initiating patrol shall tell the suspect to walk backward toward his or her voice until told to stop.

(14) The initiating patrol shall direct the suspect to the passenger side of the initiating patrol vehicle (but in front of the initial patrol vehicle's doors and between the initial and backup vehicle) into a prone handcuffing position. This procedure avoids silhouetting the assisting NSF when coming forward to take control of the suspect.

(15) The assisting NSF shall come around from his or her position and shall wait for the initiating patrol to direct the suspect to the prone handcuffing location. The assisting NSF shall handcuff the suspect and conduct a search of the rear belt area for weapons.

(16) After holstering his or her weapon, the assisting NSF shall stand the handcuffed suspect up and walk him or her back to the preferred area for a more thorough, complete search.

(17) If available, additional NSF patrols will provide assistance if needed. Any other NSF patrols on scene will maintain surveillance of the suspect vehicle.

(18) Any other occupants of the suspect vehicle should be directed to exit on the driver side and should be handled the same way as the driver.

(19) After the search is complete, the suspects shall be advised that they are being apprehended for (cite specific felony violations and other charges). It is not recommended that Miranda Act/Article 31 rights of the UCMJ be read to the suspect at this time unless other than biographical questions are to be asked. It is recommended that once it is determined who will assume jurisdiction of the investigation (NCIS/NSF), the respective interviewer/interrogator will be the one responsible to read the subject their rights as appropriate.

(20) Handcuffed and searched suspects should be placed in the primary patrol vehicle with seat belts on.

(21) Once the suspects have been secured, the vehicle should be cleared.

    (a) The assisting NSF patrol shall approach the suspect vehicle at a 45 degree angle while the initiating NSF patrol has his or her weapon at the ready with finger off the trigger.

    (b) Once the vehicle is cleared, the assisting NSF patrol should proceed back to the initiating NSF patrol and await further instructions.

(22) All suspects in the process of being checked should be searched before they are interviewed; however, a valid apprehension must precede a search for valid evidence that goes beyond a limited search for dangerous weapons.

b. Remember that circumstances of a situation may force adaptation as necessary. The idea is to safely gain control of the vehicle occupants and clear the vehicle while remaining behind cover.

E.11 ISSUANCE OF TRAFFIC TICKETS

The SO should develop RRP regarding traffic control and the issuance of traffic tickets. The RRP will include use of radar, traffic stop procedures, use of marked and unmarked vehicles, issuance of traffic tickets, distribution of the tickets, and traffic court appearance requirements. Only U.S. Armed Forces (DD 1408) and U.S. magistrate Central Violation Bureau forms will be issued. Central Violation Bureau forms will be issued for more serious violations, such as driving under the influence of drugs or alcohol, reckless driving, evading police, injury accidents, etc. NSF shall not issue state or municipal citations.

Only the U.S. Magistrate (Central Violation Bureau form) or CO/SO (for DD 1408) will negate U.S. magistrate or military traffic tickets, respectively. If the CO or SO decides that a traffic ticket should be negated due to gross administrative errors, the reason for voiding the ticket shall be written on the back of the ticket.

E.12 INTOXICATED DRIVERS

1. Driving-under-the-influence detection and deterrence measures are necessary to support prevention and assistance programs that help keep persons from harming themselves or others. Measures will include use of breath analyzers and other safety tools for detection and deterrence of impaired driving. See SECNAVINST 5300.29 (series), Alcohol Abuse, Drug Abuse and Operating Motor Vehicles, for further guidance. Breath-testing devices must be listed on the National Highway Traffic Safety Administration (NHTSA) conforming products list published in the Federal Register and shall meet current standards approved by the state or HN. Blood and urine tests may be necessary in the event of refusal of a breath test

or inability to provide breath. In the event either of these tests becomes necessary, tests will be conducted IAW Bureau of Medicine and Surgery Instruction (BUMEDINST) 6120.20 (series), Competence for Duty Examinations, Evaluations of Sobriety, and Other Bodily Views and Intrusions Performed by Medical Personnel.

2. Standardized field sobriety tests as developed by the NHTSA and other procedures for processing suspected intoxicated drivers will be developed by the SO and published via RRP. Observations of LE personnel shall be recorded on the Alcohol Influence Report (DD Form 1920). If the state or HN has not established procedures for the use of breath-testing devices, the following procedures will apply:

   a. Portable breath-testing devices shall be used during the initial traffic stop as a field sobriety—testing technique, along with other field sobriety—testing techniques, to determine whether further testing is needed on a nonportable evidentiary breath-testing device.

   b. The portable breath-testing device must be listed on the NHTSA conforming-products list and used IAW manufacturer's operating instructions. NSF must ensure manufacturer and state/local rules governing calibration requirements are adhered to.

3. The appropriate blood alcohol content level for intoxication will be based on state requirements. For installations in host nations, the applicable blood alcohol content limit is the blood alcohol content limit specified in the UCMJ or such lower limit as the SecDef may by regulation prescribe. Each command will develop RRP regarding driver refusal to be tested under implied consent laws.

4. A person being tested does not have the right to have an attorney present before stating whether he or she will submit to a test or during the actual test.

5. NSF personnel will ensure that the vehicle operated by the suspected intoxicated driver is properly secured at the scene of the apprehension, turned over to a sober driver with the authorization of the owner, or towed from the scene.

E.13 IMPOUNDING VEHICLES

1. Commanders shall ensure that policies for the impoundment and storage of vehicles are clearly defined by command policy and publicized so that operators of motor vehicles on base are able to ascertain these guidelines. The guidelines will include a description of what constitutes an unreasonable period for a privately owned vehicle to be parked illegally, when it can be impounded, and the period of time or time frame a vehicle is unattended before it is considered abandoned. Privately owned vehicle (POV) registration forms will contain or have appended to them a certificate with the following statement:

> I am aware that OPNAVINST 5530.14 (series), Navy Physical Security and Law Enforcement Program, and the installation traffic code provide for the removal and temporary impoundment of privately owned motor vehicles that are either parked illegally for unreasonable periods, interfering with military operations, creating a safety hazard, disabled by incident, left unattended in a restricted or controlled area, or abandoned. I agree to reimburse the United States for the cost of towing and storage should my motor vehicle(s), because of such circumstances, be removed and impounded. I further understand that, IAW OPNAVINST 11200.5 (series), Motor Vehicle Traffic Supervision, the commanding officer may suspend my on-base driving privileges if my vehicle is parked in an area where parking is prohibited as a force protection measure.

2. The SO shall prepare RRP specifying the impoundment procedures to be used by the NSF. Included will be the procedures to be followed prior to impoundment (e.g., notification of owner), procedures used for the actual impoundment (method of transportation, stowage, and inventories), required documentation, protection of the vehicle while impounded, and release/disposal of the vehicle. Procedures shall include a

custodial inventory, with all valuable items in the vehicle listed, as well as any notable damage, using the DON Vehicle Report (OPNAV Form 5527/12).

3. An impound record system shall be maintained that contains the date/time of impoundment, description of vehicle, vehicle ID number, license/base decal number, and final disposition, including date and to whom the vehicle was released.

4. All vehicles impounded shall be visually accounted for by an NSF representative at least monthly, regardless of whether vehicles are in the custody of a contract wrecker service or installation impound lot. This accountability will be documented in the impound record system.

5. When vehicles are impounded because of the death of the owner, close coordination with the SJA and the supply officer are required.

APPENDIX F

Jurisdictional Authority

F.1 GENERAL

The installation SJA is the focal point for determining jurisdiction. NSF must know the jurisdiction on the installation, including the geographic boundaries and the types of federal jurisdiction pertaining to AORs (i.e., exclusive, concurrent, proprietary), and the persons subject to the authority of the NSF, the UCMJ, the MCM, 18 U.S.C. 1382, citizen's arrest, the Posse Comitatus Act (18 U.S.C. 1385), SOFAs, and any agreements or understandings with local LE or HNs.

The SO will describe the jurisdiction of the NSF within written RRP and ensure they understand their jurisdictional limits. RRP must include, and all NSF must be familiar with, the geographic and subject boundaries of their jurisdiction, any agreements or understandings with local LE, and any formal agreements with HNs. The local SJA will assist the SO in drafting instructions, training, and RRP pertaining to authority and jurisdiction.

F.2 JURISDICTION CATEGORIES

Jurisdiction includes the authority to investigate and prosecute offenses. Jurisdiction depends on the nature of the offense and where the offense occurs. There are three categories of federal territorial jurisdiction. The LE-related implications of each are discussed below.

F.2.1 Exclusive Jurisdiction

The federal government may investigate and prosecute violations of federal laws as well as, under the Assimilative Crimes Act (18 U.S.C. 13), violations of state law that occur on the property. The state has no authority to investigate or prosecute violations of state law that occur onboard the installation.

F.2.2 Concurrent Jurisdiction

The federal and state governments each have the right to prosecute and investigate violations of federal and state law, respectively. In addition, the federal government may prosecute violations of state law under the Assimilative Crimes Act (18 U.S.C. 13).

F.2.3 Proprietary Jurisdiction

The federal government has the authority to investigate and prosecute nonterritory-based federal offenses committed on the property, such as assault on a federal officer, but not violations of state laws. The state has the authority to investigate and prosecute violations of state laws that occur in the territory.

F.3 NAVY SECURITY FORCES RESPONSE

NSF will respond to any criminal or other emergency incident occurring aboard the installation. NSF may respond to off-base incidents only in those cases where there is a clearly established U.S. government interest, such as military aircraft mishaps. Preliminary response to incidents includes efforts to prevent loss of life or mitigate property damage and to contain or isolate any threat to safety as necessary. After the NSF preliminary response, many incidents may call for referral to other agencies for their subsequent assumption of jurisdiction. Other agencies that may assume control and jurisdiction include the FBI, NCIS, Environmental Protection Agency, EOD, and the fire department.

Within DON, NCIS is primarily responsible for investigating actual, suspected, or alleged major criminal offenses. Major criminal offenses include offenses punishable by death or imprisonment for a term exceeding one year. The NSF shall immediately notify NCIS regarding major criminal offenses before any substantive investigative steps are taken, including interrogations or searches of property, unless such steps are necessary to protect life or property or to prevent the destruction of evidence.

1. In instances where an immediate response by NCIS is not feasible, COs may conduct such preliminary investigations as circumstances dictate, preparatory to a later full investigation by NCIS. NCIS should be immediately notified to facilitate guidance to the command.

2. NCIS may decline the investigation of some cases. If this should occur, the requesting command will document on the IR that the case was declined by NCIS, including the time/date notified, agent notified, the time/date declined, the agent declining, and reason investigation was declined.

3. Additional guidance concerning coordination with NCIS may be found in SECNAVINST 5430.107, Mission and Functions of the Naval Criminal Investigative Service.

4. NCIS is responsible for liaison with all federal LE, security, and intelligence agencies and for primary liaison with state and local agencies in matters of criminal investigation and counterintelligence.

    a. This principle does not preclude the SO from coordinating directly with the police on matters of mutual aid or other local agreements.

    b. NSF may support civilian LE agencies as authorized by federal laws and executive orders.

    c. NSF may provide support to the U.S. Secret Service (USSS) and U.S. State Department. Refer requests for assistance from other federal agencies to NCIS. When appropriately tasked to assist, NSF support the USSS in the protection of POTUS and Vice President of the United States, major political candidates, and visiting foreign heads of state. When assigned to such duty, NSF are subject to the overall supervision of the Director, USSS, or Director of Diplomatic Security, as appropriate.

5. When authorized by a commander in support of a military purpose, NSF may apprehend a military member off base in keeping with the limitations of the MCM Rule 302. A suspected military member in civilian attire should first be identified by a civilian LE officer to avoid the unlawful apprehension or detention of a civilian.

F.4 AUTHORITY

1. The authority of NSF personnel to enforce military law, orders, and regulations is derived primarily from the powers granted by Congress and POTUS. Article 7 of the UCMJ and MCM Rule 302 are the basis upon which these personnel are empowered to apprehend military offenders.

2. While there is no statutory authority granted to apprehend civilians who are not subject to the UCMJ, NSF personnel, under the authority of the CO and citizen arrest principles, may detain an individual suspected of an offense IAW the provisions of U.S. Navy Regulations, Article 0809. Title 18 U.S.C. and the Assimilative Crimes Act (18 U.S.C. 13) serve as the basis for the authority over civilian offenders.

3. Unless specifically permitted by the U.S. Constitution or by statute, the Posse Comitatus Act (18 U.S.C. 1385) prohibits direct use of the Army and Air Force in civilian LE. The SecDef and SECNAV have applied the same restrictions to DON.

    a. Courts have construed the Posse Comitatus Act (18 U.S.C. 1385) to apply to civilian NSF personnel. Courts have relied on the military purpose exception to approve military assistance in LE activities where military personnel perpetrated the illegal acts, where civilians committed offenses on military bases, or where the offense has some other military connection. See SECNAVINST 5820.7 (series), Cooperation with Civilian Law Enforcement Officials.

b. The Posse Comitatus Act (18 U.S.C. 1385) does not prevent military authorities from taking action on incidents involving civilian personnel when such action involves a specific military purpose. It allows a military member acting in an unofficial capacity to make a citizen's arrest or to take other action to preserve the public peace. It does not preclude the NSF from using force to stop a fleeing felon or suspected felon for the purpose of aiding civilian LE on a military installation.

4. Through RRP and training, the SO should ensure NSF understand the extent of their authority. Authority is the legal right to exercise power or control over individuals.

5. NSF may exercise authority over persons subject to the UCMJ. All active duty personnel are subject to the UCMJ, as well as midshipmen, reservists on inactive-duty training, and some retired members as outlined in Article 2, UCMJ.

6. The term martial law means "the temporary military government of a civilian population." Declaring U.S. federal martial law might require the United States to exercise jurisdiction over the civilian population. In time of an emergency, military jurisdiction over the civilian population extends beyond the restoration of law and order. It provides relief and rehabilitation of the people, the resumption of industrial production, the reestablishment of the economy, and the protection of life and property. With regard to personal liability, military and civilian courts may review acts performed by military personnel during martial law for damages or in criminal proceedings.

INTENTIONALLY BLANK

APPENDIX G

Crime Scenes

G.1 GENERAL

No steadfast rule can be applied to defining the size or limits of a crime scene. Areas that are well removed from the actual crime scene frequently yield important evidence. The scene of a crime is, in itself, evidence. Valuable physical evidence is normally found at or near the site where the most critical action was taken by the criminal against the victim or property. Just as it is likely to discover and develop critical evidence in the immediate area surrounding the body in a homicide, the site of forcible entry into a building often has the greatest potential for yielding evidence. Technological advances in forensic science, coupled with the ability of crime laboratories to successfully analyze evidence, have greatly increased the investigator's responsibility. Items not previously thought to be of evidentiary value may hold the key to the successful identification or elimination of a suspect or lead to the identification and conviction of the offenders. The responding patrol is responsible for securing the crime scene to support the investigation effort.

G.2 LEGAL AND SCIENTIFIC REQUIREMENTS CONSIDERATIONS

The legal requirements concerning evidence include each piece of evidence, its exact location, where and when the item was collected, and maintenance of a proper chain of custody throughout the process. Investigators' notes, photographs, and sketches become the legal record of their actions and aid immeasurably with introducing physical evidence in court. The written and photographic records of the scene also put the presented evidence into perspective and help laboratory experts determine what tests are needed. Finally, the investigators' notes, photographs, and sketches provide a valuable reference of their actions during the processing of the crime scene.

Scientific requirements in handling and processing evidence include precautions that must also be met to minimize any change or modification to the evidence. Biological materials always undergo some change. Weather and other unavoidable circumstances may induce change in some types of materials. The use of clean, spill-proof containers and proper packing materials reduces spillage, evaporation, or seepage of a sample or alteration of an item by accidentally scratching or bending it.

G.3 INITIAL RESPONSE

It is important for the first responder at the scene to be observant when approaching, entering, and exiting a crime scene. The initial responder must engage every sense to detect and determine the presence of potential threats, such as people, chemicals, or explosive devices. One of the most important aspects of securing the crime scene is to preserve the scene with minimal contamination and disturbance of physical evidence. The initial response to an incident should be expeditious and methodical. Upon arrival, responders should assess the scene and treat the incident as a crime scene:

1. When approaching and entering a crime scene, remain observant of any people, vehicles, events, potential evidence, and environmental conditions.

2. Note or log dispatch information, such as the address or location, time, date, type of call, and parties involved.

3. Be aware of any individuals or vehicles leaving the crime scene.

4. Approach the scene cautiously, scan the entire area to thoroughly assess the scene, and note any possible secondary crime scenes. Look for any individuals and vehicles in the vicinity that may be related to the crime.

5. Make initial observations (look, listen, and smell) to assess the scene and ensure safety precautions are taken before proceeding.

6. Remain alert and attentive. Assume that the crime is ongoing until it is determined otherwise.

7. Treat the location as a crime scene until it has been assessed and determined to be otherwise.

G.4 SAFETY PROCEDURES

The safety and physical well-being of NSF and other individuals, in and around the crime scene, is the first priority of the initial responding patrol. The control of physical threats will ensure the safety of the NSF and others that are present. Patrols arriving at the scene should identify and control any dangerous situations or individuals. They should:

1. Ensure that there is no immediate threat to other responders by scanning the area for sights, sounds, and smells that may be dangerous to personnel, such as HAZMAT, gasoline, and natural gas. If the situation involves a clandestine drug laboratory, biological weapons, or weapons of mass destruction (WMD), contact the appropriate personnel or agency. Before entering a possibly contaminated space or building, have HAZMAT personnel ensure the area is safe.

2. Approach the scene in a manner designed to reduce the risk of harm to other NSF patrols while maximizing the safety of victims, witnesses, and other individuals in the area.

3. Survey the scene for dangerous individuals and control the situation.

4. Notify dispatch as well as other patrols and call for backup and other appropriate assistance as needed.

G.5 EMERGENCY CARE

After controlling any dangerous situations or individuals, the next responsibility of the first responding patrol is to ensure that medical attention is provided to injured individuals while minimizing contamination of the scene. Assisting and guiding medical personnel during the care and removal of injured individuals will diminish the risk of contamination and loss of evidence. The first responding patrol should ensure that medical attention is provided with minimal contamination of the scene. The patrol should:

1. Assess the victim(s) for signs of life and medical needs and provide immediate medical attention.

2. Call for medical personnel and guide them to the victim to minimize contamination or alteration of the crime scene.

3. Point out potential physical evidence and instruct responders to minimize contact with the evidence. For example, ensure that medical personnel preserve all clothing and personal effects without cutting through bullet holes or knife tears, and document any movement of individuals.

4. Instruct personnel not to clean up the scene and not to remove or alter items originating from the scene.

5. Obtain the name, unit, and telephone number of attending personnel and the name and location of the medical facility where the victim was taken if medical personnel arrived first.

6. Document any statements or comments made by the victims, suspects, or witnesses at the scene.

7. Request that a patrol go with the victim or suspect to document any comments made and preserve evidence if the victim or suspect is transported to a medical facility.

G.6 CRIME SCENE SECURITY

Controlling, identifying, and removing individuals at the crime scene are important functions of the initial response to the crime scene. Controlling the movement and limiting the number of persons who enter the crime scene are essential in maintaining scene integrity, safeguarding evidence, and minimizing contamination. To protect the scene:

1. Establish an on-scene command post.

2. Restrict individuals' movement and activity to maintain safety at the scene and prevent them from altering or destroying physical evidence.

3. Identify all individuals at the scene.

4. Secure suspects and separate them from one another.

5. Secure and separate witnesses from one another to prevent collaboration of their stories.

6. Determine whether bystanders are also witnesses. If so, treat them as stated above; if not, remove them from the scene.

7. Control victims, family, and friends, while showing compassion.

8. Exclude all unauthorized and nonessential personnel from the scene, including all patrols not working the case, military personnel, media, and so forth.

G.7 BOUNDARY ESTABLISHMENT

Identifying perimeters assists in the protection of the scene. Establishing the boundaries is a critical aspect in controlling the integrity of evidentiary material. The perimeters of a crime scene may be easily defined when an offense is committed within a building. Regardless of the ease in which perimeters may be established, this is a measure that must be determined at the beginning of the investigation. The location and types of crime determine the number of crime scenes and the boundaries. Boundaries are established beyond the initial scope of the crime scene with the understanding that the boundaries can be reduced in size when deemed necessary but cannot be easily expanded.

1. Set up physical barriers, including ropes, cones, crime scene barrier tape, available vehicles, personnel, or other equipment. Existing boundaries such as doors, walls, and gates can be used.

2. Document the entry and exit of all individuals entering and leaving the scene once the boundaries have been established. Advise dispatch or maintain a log of each individual's name, rank, Social Security number, unit, and telephone number.

3. Implement measures to preserve and protect evidence that may be lost or compromised due to the elements (rain, snow, or wind), footsteps, tire tracks, sprinklers, and so forth.

4. Document the original location of the victim or objects that are being moved. If possible, insist that the victim and objects remain in place until the arrival of the investigation team.

5. Consider search and seizure issues to determine the necessity of obtaining consent to searching and/or obtaining a search warrant.

6. Maintain security of the scene until formally relieved by investigators or the incident commander.

G.8 DOCUMENTATION

To preserve information, the first responding patrol must produce clear, concise, documented notes attesting to actions and observations at the crime scene as soon as possible after the event occurs. These notes are vital in providing information to substantiate investigative considerations. All notes must be maintained as part of the permanent case. The patrol should document:

1. The appearance and conditions of the crime scene upon arrival, such as smells, ice, liquids, movable furniture, weather, temperature, or personal items; whether the lights are on or off; whether the shades are up, down, open, or closed; or whether the doors and windows are open or closed.

2. Personal information, statements, or comments from witnesses, victims, and suspects.

3. The actions of others and any of his or her own actions.

4. When possible (after the scene is secure), take photographs/video of the scene and provide to investigators.

Once the investigation team arrives to take control of the scene, the first responding patrol should brief the investigator in charge, relinquish responsibility for the scene, and then assist the investigation team as necessary.

Be prepared to participate in a crime scene debriefing by investigators. The crime scene debriefing is the best opportunity for LE personnel and other responders to ensure that the crime scene investigation is complete and to share information regarding particular scene findings before releasing the scene. It provides an opportunity for input regarding a follow-up investigation, special requests for assistance, and the establishment of post-scene responsibilities.

G.9 EVIDENCE-HANDLING AND CUSTODY PROCEDURES

Refer to Appendix I of this NTTP for evidence-handling and custody procedures.

APPENDIX H

Inspections and Searches

H.1 INTRODUCTION

A search is an examination of a person, property, or premise to uncover evidence of a crime or criminal intent (e.g., stolen goods, burglary tools, weapons, etc.). Search and seizure includes the procedures to obtain authorization for searches; documentation requirements for authorization and evidence; determination of the scope of a search and articles subject to search and seizure; suspect rights; rights to privacy; body searches; conducting entry; identification, automobile, area, and consent searches; search procedures; safety and security measures; and chain of custody. NSF may conduct searches of persons, property, or areas within jurisdictional limitations. General information on search and seizure is found in the MCM. Up-to-date information may be obtained from the SJA or the servicing region legal service office. With few exceptions, evidence obtained in an illegal search is inadmissible at a court-martial.

H.2 SEARCH TYPES AND CONSIDERATIONS

H.2.1 Command-authorized Searches

Command-authorized searches (probable-cause search authorizations) will be conducted only after a determination that probable cause exists that a person, property, or evidence connected to a crime is located in a specific place or on a specific person who committed the offense, and that the fruits or instruments of the offense are in a specific place. Confer with the command SJA or legal office for assistance with obtaining a command search authorization.

1. Prior to issuance of a command-authorized search, the Affidavit for Search Authorization (OPNAV 5580/10) shall be completed and signed, under oath.

2. The Command Authorization for Search and Seizure (OPNAV 5580/9) will be prepared for the commanding officer's signature.

3. When exigent circumstances exist and there is no time to get a search authorization without substantial risk of loss of evidence, escape of individuals, or harm to innocent people, the warrant or command authorization requirement may be excused; however, probable cause must still exist.

4. In conducting the search, the individual(s) having proprietary interest over the premises should be present. A copy of the signed OPNAV 5580/9 must be given to the individual, and he or she will be given sufficient time to read it. If the individual is also the suspect of the offense, then no questions other than requests for identification shall be asked without appropriate self-incrimination warnings. After serving the authorization, NSF may ask the individuals present to open locked doors, lockers, etc., providing they are cooperative and not a risk to the safety of security personnel or likely to damage or conceal evidence. NSF members shall not ask the suspect where the item(s) that are the subject of the search are located. If individuals are not cooperative or refuse to open locks, the locks may be forced open by NSF in such a manner as to cause the least amount of damage to the property. Keep in mind, when justified, the manner and extent of the search are commensurate with the reason for the search.

5. Occasionally, a search authorization may be served on an unoccupied premise, vehicle, shipping container, etc. In such situations, a command representative should be present to witness the search. A copy of the authorization shall be left with the representative to be delivered to the suspect. Another copy, plus a

receipt for property seized, shall be left at the premises. It is the security personnel's responsibility to secure the unoccupied premise after the search to prevent theft and vandalism.

6. Following the search, a receipt for the property seized must be completed and provided to the individual with proprietary interest in the property, along with a copy of the authorization. A signed receipt must then be returned to the commanding officer authorizing the search. The original of the search authorization and affidavit is retained by the security detachment conducting the search.

H.2.2 Consent Searches

1. A Permissive Authorization for Search and Seizure (OPNAV 5580/16) shall be completed prior to any consent search. Oral consent authorization may be made in emergency situations, but only with sufficient witnesses present to testify that permission for the search was voluntary and not mere acquiescence to authority.

2. NSF must ensure that the party who consents has cognizance over the area to be searched and is not asked to identify property or an area to be searched, as such identification might be an unwarned admission, causing evidence discovered to be tainted and therefore inadmissible.

3. NSF should advise the subject that the permission to search is voluntary. NSF should make no further statement to the individual to imply that if he or she does not consent, a search warrant will be obtained and the search conducted regardless. NSF should record the exact words of a subject and of the person giving the advice in NSF field notes.

4. If, during the course of a search, subjects state they do not want the search to continue, then NSF shall terminate the search. Any evidence uncovered to that point may be used to substantiate probable cause, and the matter presented to the commanding officer for a search authorization, if appropriate.

5. At the completion of the search, NSF shall provide to the subject a receipt for any property seized. There is no requirement to provide a copy of the permission authorization form to the subject, although NSF may do so if the subject requests it. The original form shall be kept by the security detachment conducting the search. The last page, listing only Blocks 1–16 of the Evidence/Property Custody Document (OPNAV 5580/22), shall be used as a receipt for property seized.

H.2.3 Other Reasons for Conducting Searches

1. Search incident to apprehension. NSF may conduct a search incident to an apprehension without obtaining search authority, and this may include the immediate area over which the apprehended person exercises control. Search incident to an apprehension should be conducted immediately.

2. Patrol safety. When NSF conducts a lawful stop, individuals stopped may be searched for concealed weapons or dangerous objects when the patrolman feels his or her safety may be in jeopardy or when it is reasonably believed that the individual may be armed and presently dangerous. Contraband or evidence located in the process of a lawful search may be seized.

3. Other types of searches are located in the MCM, Rules 314–315. Due to case law, changes occur quite often; therefore, these searches are not listed in this NTTP. Ensure the MCM is current.

H.2.4 Special Search Situations

1. Medical aid. NSF may conduct a search of persons or property in a good-faith effort to render immediate medical aid, to obtain information that will assist in the rendering of such aid, or to prevent immediate or ongoing personal injury.

2. Search of and by the opposite gender. Searching members of the opposite gender and premises occupied by members of the opposite gender is sensitive. Take certain precautions, carefully consider actions, and use common sense.

   a. NSF may search outer garments (e.g., jackets, coats, etc.) and hand-carried items of a member of the opposite gender.

   b. Regardless of the gender of the person being searched, conduct searches in the same manner. NSF or other military persons of the same gender conduct the search unless an urgent safety or security need exists.

   c. NSF may conduct a search of premises exclusively occupied by members of the opposite gender. It is recommended but not required that NSF or military personnel of the same gender as the occupants of the premises be present during the search.

   d. When possible, NSF should avoid conducting searches of personnel of the opposite gender. However, if such searches are required, ensure there is at least one witness present.

H.3 LEGAL ASPECTS

1. Search area. NSF may conduct a search for weapons or destructible evidence in the area within the immediate control of a person who has been apprehended. The area within the person's immediate control is the area that the individual searching could reasonably believe the person apprehended could reach with a sudden movement to obtain such property.

2. Entry to U.S. property. Inspections upon entry to U.S. installations, aircraft, and vessels abroad do not require probable cause or consent to search.

3. Installation ECP inspections and searches. The CO may order NSF to inspect all or a percentage of motor vehicles entering or leaving the installation. The CO may also authorize searches of specific motor vehicles in the same manner as premise searches.

4. Off-installation searches. Comply with local, state, and federal law when conducting off-installation searches of persons subject to the UCMJ or their property. The CO must review and approve any requests for such searches. Seek the advice of the installation SJA prior to any search.

5. Searches outside the United States, U.S. commonwealths, and U.S. territories.

   a. Authority for conducting search and seizure operations outside U.S. federal jurisdiction varies according to geographic locations and U.S.-HN agreements. Consult with the local SJA before conducting off-installation searches outside the United States, U.S. commonwealths, and U.S. territories.

   b. Searches conducted by foreign nationals. Commands may not delegate to a foreign national the general authority to order or conduct searches. When making a lawful apprehension, HN contract police may search the suspect's person and clothing worn, and the property in the suspect's immediate possession. HN contract security police may also search a motor vehicle that a suspect was operating or riding in as a passenger. HN law or U.S.-HN agreements govern other restrictions or authorizations.

H.4 EVIDENCE PROCUREMENT AND HANDLING

NSF who are lawfully in any place may, without obtaining a warrant, consent, or commander's authorization, seize any item in plain view that they have probable cause to believe is contraband or evidence of a crime. This is so even if the item in question is not related to the crime currently being investigated.

1. The following are items that may be seized during a lawful search:

    a. Items obtained during a crime.

    b. Items from another crime even if the second crime was never reported or its existence is otherwise unknown to them.

    c. Instruments used to commit a crime.

    d. Contraband—material the possession of which is by its very nature unlawful. Material may be declared to be unlawful by appropriate statute, regulation, or order. See MCM, Appendix 22, Military Rule of Evidence 313(b), pp. 22–23.

    e. Weapon—any item that is or can be used as a weapon may be seized for the protection of LE personnel, even if the item is not connected with the offense for which the search was authorized.

    f. Evidence of a crime—e.g., blood-stained clothing, soil samples from shoes, or unique items of clothing or jewelry that were previously described by a witness.

2. Searches must comply with the limitations imposed by the authorization. Only those locations described in the authorization may be searched, and the search may be conducted only in areas where it is likely that the object of the search will be found.

3. If the authorization for search is for a specific item or quantity of items, such as a stereo receiver by serial number or clothing by description, then NSF must terminate the search as soon as the property is located. If the search authorization is for a nonspecific quantity or class of items, such as narcotics or financial records, then NSF may continue the search throughout the premises since there is no way of determining how many of these items are present.

4. Evidence is to be documented.

    a. Record all circumstances surrounding the discovery of evidence (e.g., location of the discovery, date and time, witnesses present, etc.). These notes provide facts for an incident report. Additionally, NSF may use these notes to testify in court. As a minimum, file a copy of the notes with the IR and retain original notes.

    b. Place evidence tags—Evidence/Property Custody Document (OPNAV 5580/22)—on all items that have evidentiary value. Use care not to destroy the evidentiary value of the item through the careless marking of the item. Use envelopes, boxes, plastic bags, etc., to collect evidence. Exercise sound judgment to avoid damaging a valuable stolen item that may eventually be returned to its owner.

    c. Maintain a complete chain-of-custody accounting of all personnel who handle evidence.

    d. Return all evidence items to their rightful owners upon final disposition of a case. Coordinate all releases of evidence with the SJA.

H.5 CONDUCT OF PERSONNEL SEARCHES

1. Search of suspects. NSF searches suspects for weapons or evidence. All apprehended suspects shall undergo a search prior to transport to ensure the safety of both NSF and suspects. The apprehending NSF member makes the decision to search. A search may escalate if NSF members detect an object that they believe may be a weapon or other instrument likely to cause death or bodily harm. At that time, NSF may begin an intrusive search into the garment, etc., to retrieve the item. Base this decision on the situation. Searching is potentially a dangerous time for NSF and, if not properly conducted, may result in serious bodily injury or death. Search techniques exist to minimize these dangers. NSF who fail to take full advantage of these techniques pose a threat to themselves and their fellow NSF members. A good rule of thumb is to remember that all apprehended suspects are potentially dangerous; NSF should afford themselves all possible protection.

2. Patrol safety. Search apprehended suspects as soon as possible after apprehension unless the situation dictates otherwise. NSF members should not search an individual without the aid of another NSF member. The assisting NSF member participating in the search provides additional suspect control and reduces the probability of the suspect's escalating the risk. For MWD teams, the dog is considered the assisting member and is positioned as required. Do not draw a weapon unless its use is imminent or the reason for apprehension would justify its use.

3. Decision to search. NSF must make an on-the-scene decision and assessment of each situation to decide which search method to employ. No matter which one is used, conduct it quickly and thoroughly. For quickness, search systematically; for thoroughness, never pat or run hands over the suspect lightly—grab or squeeze every inch of the suspect's clothing during the search. Although a NSF member can start a search from either side, it should always start from the same side the hand rotation technique is applied to maintain effective control over the suspect.

4. Precautionary check. Make a precautionary check of the suspect prior to the actual search. The precautionary check will consist of verbal communication with the suspect. Ask suspects if they have any sharp objects on their person (e.g., needles, razor blades, or knives). Regardless of the suspect's reply, proceed with caution to determine the location of any objects. Remember that this is a high-risk environment for health hazards such as exposure to HIV, Hepatitis B virus, or other blood-borne pathogens resulting from a stick or cut.

5. Types of searches. The two most commonly used searches by NSF are the standing and prone searches. Apply handcuffs before beginning any search.

   a. Standing search. The standing search is used primarily on suspects who do not appear dangerous or violent or are not so drunk/drugged that they cannot stand under their own power.

      (1) With one patrol member at the suspect's right rear, the assisting NSF member should take an over-watch position in front of the suspect, opposite the side of the searching member, to maintain surveillance of the search procedures. The assistant must remain close enough to physically aid the searching NSF member if necessary.

      (2) As the searching member, instruct the suspect to turn his/her head facing the opposite direction of intended approach. Next, grab the handcuff link chain with the hand palm down. Keep your weapon or strong side away from the suspect while placing your foot directly behind and centered between the suspect's feet. Instruct the suspect to keep his/her head up and knees flexed. If the suspect becomes uncooperative or aggressive, control is gained by pulling the handcuffs down and back to keep the suspect off balance (figure H-1).

      (3) Search the suspect's entire side from the position you are in. If the suspect is wearing a hat, remove it and search it: Carefully check the hat for sharp objects, weapons, and contraband. Pay particular attention to the area beneath the sweatband and seams. After searching the hat, drop it on the ground behind the suspect. Use the hat as a container for any items taken from the suspect.

      (4) Run your fingers through the suspect's hair and around the left side of the head. Then search the back of the suspect's neck (figure H-2).

      (5) Work down the side of the suspect's back, checking across the spine to the left side and down from the collar to the waist (figure H-3).

      (6) Search the suspect's side to the armpit and down the underside of the arm to the suspect's hand.

      (7) Check the rest of the suspect's arm from the hand to the shoulder. Search the suspect's throat and chest working down to the waist (figure H-4).

Figure H-1. Standing Search

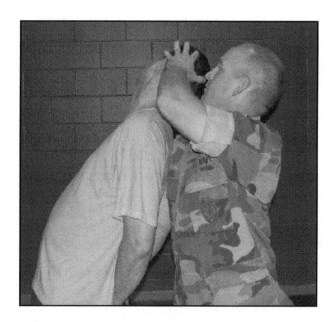

Figure H-2. Standing Search (Neck)

Figure H-3. Standing Search (Back)

Figure H-4. Standing Search (Chest)

(8) Search the suspect's side to the armpit and down the underside of the arm to the suspect's hand.

(9) Check the rest of the suspect's arm from the hand to the shoulder. Search the suspect's throat and chest working down to the waist (figure H-4).

(10) Starting at the suspect's waist, in front, search the waistline to the small of the back (figure H-5).

(11) Move the left foot back for balance and crouch down. Pull the handcuffs downward and back to maintain control.

(12) Search the suspect's buttocks and groin area. Search down the inside of the suspect's left leg, inside the top of the left shoe, under the arch of the shoe, and back up the outside of the suspect's left leg to the waist (figures H-6, H-7, and H-8).

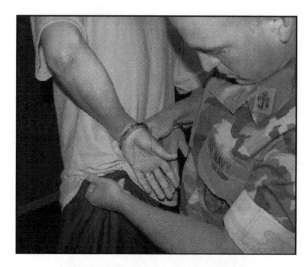

Figure H-5. Standing Search (Waist)

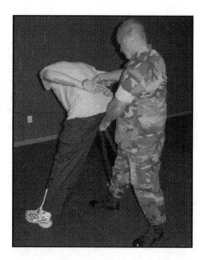

Figure H-6. Standing Search (Buttocks)

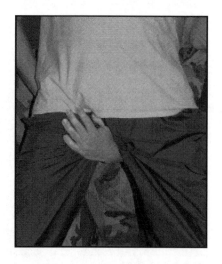

Figure H-7. Standing Search (Groin)

Figure H-8. Standing Search (Leg)

(13) After searching the suspect's left side, assume the initial control position.

(14) Place the left hand over the right hand and the handcuff linking chain. This step is commonly known as the "crossover."

(15) Slide the right hand out and grip the linking chain with the left hand (figure H-9).

(16) Turn so your left side is toward the suspect's back and your right side is at a 45 degree angle from the suspect. Place the left foot directly behind and centered between the suspect's feet. Again, direct the suspect to look away from the side being searched (figure H-10).

(17) Search the suspect's right side using the same procedures used for the left side. During any part of the search, if the suspect violently resists, pull the handcuffs down and back to maintain control.

b. Prone search. The prone search is used primarily on suspects who appear to be dangerous or violent or are so drunk/drugged that they cannot stand under their own power.

(1) Place the suspect in the prone position with his/her hands cuffed behind his/her back and feet spread wide apart.

(2) Direct the suspect to keep his/her head turned away from the search side at all times.

(3) The assisting member is positioned approximately 6 feet from the suspect's head in the over-watch position. The assisting member will position him/herself over the suspect and opposite the side of the searching member.

(4) The searching NSF member positions him/herself over the suspect on the right side of the handcuffed suspect and searches the suspect's left side.

(5) The searching member will place his/her left knee on the suspect's right leg at the knee area and keeps his/her right foot flat on the ground next to the suspect's side (figure H-11).

(6) If the suspect attempts to struggle, the searching member will drop his/her right knee onto the suspect's back. Grab the suspect's hand and bend the palm into the wrist while pulling the handcuffs toward the suspect's head. Continue to apply pressure compliance techniques until control is regained.

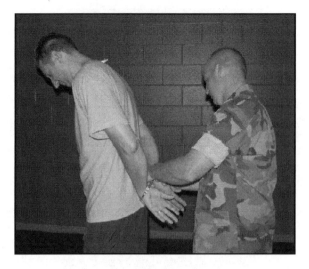

Figure H-9. Standing Search (Handcuffs)

Figure H-10. Standing Search (Control)

Figure H-11. Prone Search

(7) If the suspect is wearing a hat, search the hat before beginning the full body search. After the hat is searched, use it as a container for items taken from the suspect. With the left hand, search the outer half of the suspect's left leg from the waist down to the left knee (Figure H-12). Lay the palm down on the suspect's knee and instruct him/her to raise his/her left leg (Figure H-13). Use the palm to block a possible kick as the suspect raises his/her leg.

(8) Search the outer leg to the foot, then the shoe, and back up the inner leg. After completing the inner leg to the knee, have the suspect lower his/her foot to the ground. Search up the inner leg to the groin and left buttock (Figure H-14).

(9) Place the right arm in the crook of the suspect's left arm and roll him/her back until his/her left side is slightly off the surface. Lean the right knee against the suspect's back for balance (Figure H-15).

(10) With the left hand, release the linking chain and search the suspect's waistband from the middle of the back to the belt buckle (Figure H-16).

(11) Search the suspect's lower left abdomen and groin, then up the front side of the suspect's torso to the throat (Figure H-17).

(12) Return the suspect to the prone search position.

(13) Cross the right hand over the left hand. Place the right hand in the small of the suspect's back while keeping the left hand on the linking chain of the handcuffs (Figure H-18).

(14) Keep pressure on the suspect's back to maintain control. Rise up and step over the suspect's buttocks (Figure H-19).

(15) Rest the right knee on the suspect's left leg just above the knee and the left foot flat on the ground against the suspect's right side (Figure H-20).

(16) Instruct the suspect to turn his/her head away from the side to be searched. Search the suspect's right side using the same procedures/techniques outlined for searching the suspect's left side. Note that the above search procedures are for right-handed individuals. Left-handed persons may start the procedures from the opposite side.

(17) After the search is completed, bring the suspect to a standing position.

(18) If the suspect is unable to stand alone, roll him/her onto his/her side.

(19) Sit the suspect on his/her buttocks.

(20) Have the suspect bend his/her left or right leg at the knee, bringing the foot to the groin.

(21) Assist the suspect to roll onto the kneeling position.

(22) From the kneeling position, place left hand on the inside of the suspect's left shoulder and right hand on the back of right shoulder. Twist suspect counterclockwise into a standing position.

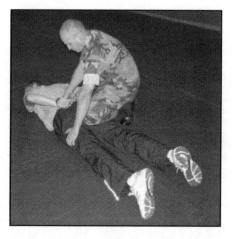

Figure H-12. Prone Search (Full Body)

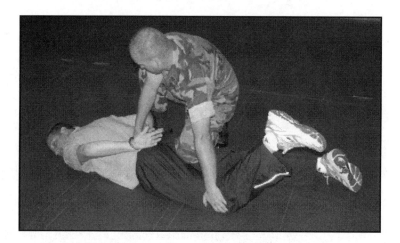

Figure H-13. Prone Search (Leg)

Figure H-14. Prone Search (Inner Leg)

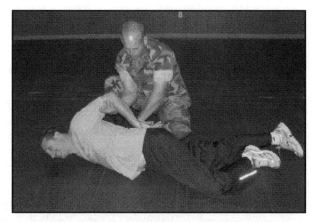

Figure H-15. Prone Search (Left Side)

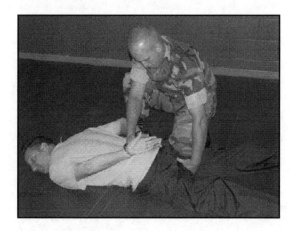

Figure H-16. Prone Search (Waist)

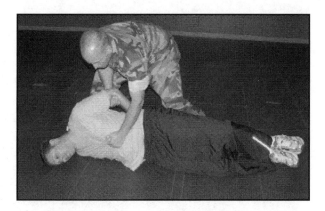

Figure H-17. Prone Search (Abdomen/Groin)

Figure H-18. Prone Search (Handcuffs)

Figure H-19. Prone Search (Back/Buttocks)

Figure H-20. Prone Search (Control)

APPENDIX I

Evidence Handling and Custody Procedures

I.1 GENERAL

NSF personnel should take every precaution to preserve the integrity of evidence in its original condition. NSF personnel must enter evidence into the evidence custody system as soon as possible after its collection, seizure, or surrender.

Host activities will receive and store evidence received from tenant activities. Tenant activities shall not operate their own evidence lockers unless they have the capacity to do so, including suitable storage containers that comply with this manual and a formal agreement with the host activity.

I.2 RESPONSIBILITIES

1. The SO shall:

   a. Establish an evidence custody system consisting of:

      (1) An evidence custodian, designated in writing

      (2) An alternate evidence custodian, designated in writing

      (3) An evidence locker/room meeting the security construction and locking requirements identified in paragraph I-4

      (4) A bound evidence log (OPNAV 5580/24), an active evidence custody file, and a final evidence disposition file.

   b. Conduct a complete evidence inventory on relief of the SO, evidence custodian, or alternate evidence custodian, whenever a discrepancy is noted, and at least semiannually.

2. The evidence custodian (or the alternate evidence custodian in the absence of the primary) shall ensure that:

   a. Evidence is inventoried, tagged, packaged, and marked prior to storage and that evidence custody documents are properly completed by the person delivering the evidence prior to its acceptance

   b. Evidence logs and records are properly maintained

   c. Evidence is disposed of following current Navy and command policies.

I.3 EVIDENCE RECORDS

1. Evidence custody records will be composed of a bound evidence log (OPNAV 5580/24), an active evidence custody file, and a final evidence disposition file.

2. The evidence log shall be maintained for a period of five years from the date of the last entry. The active evidence custody record shall be maintained as long as there is any evidence in custody that has not been officially disposed of. The final evidence disposition file shall be maintained for a period of five years after the close of the calendar year covered by the file.

3. Each custody document received by the custodian shall be reflected on a separate logbook line. The log shall also contain dated entries of all inventories; all changes of SO, evidence custodian, or alternate; and all changes of lock combinations. Each entry indicating a receipt of evidence by the custodian or alternate custodian shall be assigned an evidence log number consisting of two groups of numbers separated by a hyphen (-); the first set of numbers will be a three-digit chronological number of the document for that calendar year, and the second group will consist of the last two digits of that year (e.g., 001-08 for the first evidence custody document for the calendar year 2008). Each entry shall be made in black ink on the next blank line, and no empty lines shall be permitted. In the event an error is made in the entry, the entry shall be ruled out with a single line and initialed by the custodian. Erasures or entries obscured with correction fluid are not authorized.

4. The active evidence custody file shall consist of copies of each Evidence/Property Custody Document (OPNAV 5580/22) relating to evidence that has been received by the custodian and has not been disposed of. This record shall be maintained in one or more loose-leaf notebooks and filed by evidence log number, with newer entries being placed on top. This file will identify all evidence for which the evidence custodian is responsible.

5. A final evidence disposition file shall be maintained on all ECDs relating to evidence that has been disposed of. This file will be kept in the same manner as the active evidence custody file. When it is ready for filing, the evidence custodian shall complete the final disposition section. The duplicate copies of ECDs shall be destroyed. In the event the original ECD is forwarded with the evidence during the final disposition action, the copy in the active evidence custody record will be completed as if it were the original and transferred to the final evidence disposition file.

6. The required evidence custody records shall be stored in the evidence locker to prevent unauthorized access to the files.

7. Only approved OPNAV evidence logs and forms are to be used for evidence custody. If evidence is received from a non-Navy source, an OPNAV evidence custody document shall be initiated and the original evidence document will be attached to it.

I.4 EVIDENCE LOCKERS

1. Evidence lockers shall be used to safely secure all items of evidence. Lockers shall meet the construction and locking criteria specified for "Q" and "R" coded items; CAT IV AA&E per OPNAVINST 5530.13 (series), Department of the Navy Physical Security Instruction for Conventional Arms, Ammunition, and Explosives (AA&E).

   a. If only small items of evidence are held, they may be stored in GSA-approved security containers. However, if the container weighs less than 750 pounds, it shall be secured to the floor/wall. GSA-approved containers are not allowed to be physically modified in any fashion; therefore, containers of less than 750 pounds can be made to meet the weight criteria by adding weight to the inside of the container or by using straps across the container with the straps bolted to the floor/wall. If straps are used, bolts shall be modified to make them nonremovable.

   b. Larger items or amounts of evidence shall be stored in a vault or strong room meeting the construction requirements of SECNAVINST 5510.36 (series), Department of the Navy (DON) Information Security Program (ISP) Instruction.

2. Only GSA Group 1 or Group 1R three-tumbler combination locks shall be used as locking devices. For evidence storage, electromagnetic locks are not required. Key-operated locks and cipher locks will not be used.

3. Construction and lock requirements do not extend to temporary evidence lockers (drop boxes) that meet the following specifications:

   a. The drop box is constructed so that once evidence is placed inside, it cannot be removed without a key without destroying the box or lock.

   b. The box is within the continual sight of the dispatcher or a security supervisor.

   c. The key to the drop box is stored in the regular evidence locker.

   d. The evidence is not held in this temporary storage for more than 24 hours.

4. All evidence received by a Navy component having an evidence locker should be stored therein, with the exceptions of evidence too bulky for storage, evidence of a classified nature requiring special handling (e.g., sensitive compartmented information and communications security), perishable items such as food and human/animal parts, items of an unstable chemical/flammable nature, and explosives. With the prior authorization of the SO, these types of evidence may be stored elsewhere where restricted physical access to the evidence can be maintained. In any event all such items, unless of an especially bulky nature, will be wrapped or placed in containers and sealed so that any unauthorized access to the evidence can be detected. Personnel maintaining temporary custody of these items shall be briefed on the requirements for secure storage and the probable requirement for them to testify as to their custody. Further, they shall execute the ECD upon receipt and release of the evidence. The original ECD may be left with the evidence unless the storage conditions might cause its destruction. In this case, a copy may be substituted and the original maintained in the evidence locker. It may be necessary to destroy some extremely hazardous items that present an immediate danger to an installation or personnel. In these cases destruction shall be made and witnessed IAW established EOD or HAZMAT procedures and documented as required in paragraph I.9. Consideration should also be given to documenting such destruction via videotape and photographs.

5. Access to the evidence locker:

   a. Should be strictly limited to the evidence custodian, the alternate evidence custodian, and the SO. The SO shall be provided the combination to the locker in a sealed Security Container Information envelope (SF 700). Should the SO gain access to the evidence locker by removing the combination from the envelope, the combination will be changed and a new SF 700 will be provided to the SO. The reason for the entry by the SO will be recorded in the bound evidence log on the day it occurs.

   b. Persons other than the evidence custodian, SO, and alternate evidence custodian may be granted access for official purposes (e.g., assistance visits/inspections). Approved visitors shall be entered in a log showing the date of the visit, time, identity of the visitor, office or official capacity of the visitor, and reason for the entry into the locker. Approved visitors shall be accompanied by the evidence custodian or alternate evidence custodian while accessing the evidence locker.

6. At no time shall lost and found property, supplies, cleaning gear, or any other nonevidentiary items be placed in the evidence locker.

I.5 EVIDENCE SUBMISSION

1. When any person submits evidence to the evidence custodian, it should be properly tagged and, if appropriate, placed in a suitable container. The original and two copies of the ECD shall be securely attached to the evidence or its container. Only the custodian accepting the evidence shall separate the original and copies of the ECD and the interleaved carbon paper. The evidence custodian shall sign the

original and all copies of the ECD in the appropriate block acknowledging receipt of the evidence. The original ECD shall remain physically attached to the evidence; the first copy shall be placed in the active evidence custody file, and the second shall be returned to the person submitting the evidence. When the evidence is turned into a temporary depository (drop box), the person depositing the item shall sign the ECD in the "Released by" column and shall enter the name or number of the depository in the "Received by" column. When the seizing person is also the evidence custodian, he/she shall also complete the "Released by" column on the ECD to show release by the seizing person and receipt into the evidence custody system. When any evidence is checked out of the evidence locker for whatever purpose prior to its final disposition, a copy of the ECD shall be maintained in the evidence locker in the active evidence custody file. In the event that the original is lost or destroyed, a copy may be used in its place.

2. All evidence being submitted will be carefully examined and counted/weighed by the accepting custodian. Any items entered that are later found not having evidentiary value should be promptly returned or disposed of IAW policy established by the CO, SO, or SJA. If there is any doubt as to the value of the evidence, advice should be sought from the servicing Navy legal service office.

3. When pills/capsules are entered into the evidence locker, they shall be counted and not weighed. However, if a circumstance exists that would require weighing, then the submitter should consider using the term "approximately" somewhere in the descriptive portion of the ECD.

4. When additional evidence collected is applicable to a case already having an evidence log number, the new evidence shall use the same number with a letter suffix (e.g., 001-08A). Each new ECD pertaining to the same investigation will have the subsequent letter used as the suffix.

I.6 EVIDENCE INVENTORY

1. When evidence is first received into custody, the receiving party must first inventory the evidence personally. Subsequently, when evidence is transferred between parties, the evidence must be verified. Sealed containers marked with the date and signature of the person who collected the evidence and the evidence custodian or alternate custodian who accepted the property into evidence need not be opened during an inventory unless there is evidence of tampering or the seal is broken.

2. The contents of each evidence locker shall be inventoried semiannually (six-month intervals). An inventory shall also take place on the occasion of the relief of the evidence custodian/alternate evidence custodian or the SO and upon loss or suspected loss of evidence or when a breach of security has occurred or is suspected. If the inventory required for the replacement of the evidence custodian/alternate occurs within two months of the regularly scheduled semiannual inventory, it may be substituted for that inventory.

3. Inventories shall be accomplished by the evidence custodian, the SO, and a disinterested third party. When a requirement calls for a "disinterested party" to witness evidence inventoried or destroyed, that party shall not be someone who is assigned to the security/legal departments of any command/tenant activity. The disinterested party shall be identified by name, grade/rank/rate, and billet title. Additionally, if the relief of the custodian requires the inventory, the incumbent and the relieving custodian shall conduct it with the SO and a disinterested third party.

4. The inventory required by this instruction shall, at a minimum, consist of a reconciliation of the evidence log against the active evidence custody record and a visual accounting of each item for which there is a log entry without final disposition and an ECD in the final evidence disposition file. Evidence that is stored in sealed containers, envelopes, or other means with the evidence custody officer (or alternate) and submitting officer signatures and dates on the unbroken seal need not be opened unless there is reason to believe there has been tampering. Evidence that is stored outside the command evidence locker because of its bulk, classification, or other special nature will be sighted at each inventory. The only exception to this sighting requirement will be evidence that has been temporarily transferred to another activity and for which the ECD is properly annotated.

5. When the inventory is completed, an entry shall be made in the evidence log reflecting the reason for the inventory, who conducted the inventory, and the results. Participating individuals shall sign the log above their names and billet titles. If the inventory was due to the relief of a participant, all combinations shall be changed, and the log shall reflect this action. Any discrepancies found shall be immediately reported by the SO in writing to the commanding officer. The SO shall then initiate a full investigation and prepare a final report on the discrepancy, its suspected cause, and actions taken or recommended.

6. At commands with large amounts of active evidence, the signing of the evidence log by both the incoming and outgoing evidence custodians attesting to the fact that they have completed a visual sighting of each item and have found no discrepancies shall complete the transfer of evidence. This procedure negates the need to record the transfer of evidence between evidence custodians on each and every active ECD within the system. However, where holdings are small, completing the ECD should be the method of transferring evidence.

7. Prior to the semiannual inventory of evidence, the evidence custodian shall review the active evidence custody file with the SO to identify what evidence may be logically considered for disposal.

8. During inspections, the inspector(s) may conduct a review of evidence custody procedures by reviewing selected items of evidence, the associated documentation, and a sample number of final disposal actions. This step shall not be considered as a substitute for required inventories.

I.7 TRANSFER AND SHIPMENT OF EVIDENCE

I.7.1 Temporary Transfer

When it is necessary to transfer evidence to another agency on a temporary basis, the original ECD shall accompany the evidence. Prior to the release of the evidence, the evidence custodian shall sign both the original and duplicate ECD. The duplicate will be retained in the active evidence custody file. Those persons handling the evidence prior to its return to the evidence custodian shall complete the appropriate blocks in the accompanying original ECD. A receipt must be obtained from the receiving agency. In the event that only part of the evidence will be temporarily transferred, the original ECD shall accompany that part, with appropriate notations in the item column to the left of the transferring signature. A duplicate of the original ECD shall be reproduced and attached to the balance of the evidence maintained in the evidence locker. Upon return of the original ECD, the duplicate may be destroyed.

I.7.2 Permanent Transfer

In the event that it is necessary to transfer evidence permanently to another agency, the original ECD shall accompany the evidence. The custodian shall sign the evidence out of the evidence locker. The evidence custodian shall execute the duplicate copy in the active evidence custody record, and the agency representative receiving the evidence shall complete the final disposition portion. The appropriate disposition entry shall also be made in the bound evidence log. The duplicate copy shall then be filed in the final evidence disposition file. In the event that the receiving agency does not accept all the evidence listed on the ECD, the original form will be retained with the balance of the evidence in the evidence locker. The agency representative shall provide a receipt for that portion of the evidence taken by an appropriate entry on the form, and in return will be provided a copy of the original form. The final disposition of the evidence will not be entered in the bound evidence log until all evidence listed in the ECD has been disposed of.

I.7.3 Transfer to Other Naval Commands

When evidence is transferred to another naval command, the original ECD shall be transmitted with the evidence. In the event that only part of the evidence is transferred, the original will be forwarded and a copy attached to the balance. If it is known prior to transfer that part of the evidence will not be needed, the original ECD will be retained and a copy forwarded with the evidence. The command receiving the evidence shall continue to use the custody document attached to the evidence. The item(s) of evidence will be logged the same as any others, including the assignment of a new evidence log number. The entry in the bound evidence log will show the new number followed by the log number (in parentheses) of the originating component (e.g., 085-04 (244-05)). The new number shall also be placed on the ECD just above the original number. The receiving component shall then reproduce a copy of the ECD and place it in its active evidence custody file. The component that transferred the evidence (if all its evidence was in fact transferred) will remove its copy of the ECD from its active file, appropriately annotate the disposition portion, and place the document in the final evidence disposition file.

I.7.4 Transfer by Mail

Whenever possible, transferred evidence will be hand-delivered. Evidence that is to be mailed, in all cases, must be mailed via registered mail with a return receipt requested. The registered mail receipt and the return receipt shall be stapled to the ECD in the active evidence custody file. If the transfer is permanent, the receipts will become a permanent part of the system by inclusion in the final disposition file with the document. When evidence is prepared for mailing, it shall be double-wrapped, with the inner wrapping marked to indicate the presence of evidence. The package must be specifically addressed to the evidence custodian.

I.8 DISPOSAL OF EVIDENCE

1. Approval for the disposal of evidence shall be requested in writing to the adjudicating authority or SJA if so designated. Authority, if given, will be indicated by endorsement.

2. Evidence that was used in any court-martial action should not be disposed of until the trial and subsequent appeals have been completed. Authorization for disposal must be obtained from the trial counsel handling the trial or the judge advocate of the next senior in command if on appeal. If the evidence was used in federal, state, or other civilian court, the authorization must be obtained from the appropriate prosecuting attorney prior to disposal. When authorization for disposal is received, the evidence custodian will complete the final disposition section of the ECD, by recording the name and title of the person authorizing the disposal.

3. Any evidence that was used in any administrative process shall not be released until all appeals or reviews of the initial action are completed. Prior to disposal of such evidence, authorization shall be obtained from the judge advocate or command legal officer of the command that has cognizance over the person against whom the action was taken. In the event of their absence, their counterparts at the next senior command should be contacted. When authorization is received, the evidence custodian shall complete the final disposition section of the ECD, indicating the name and title of the person authorizing the disposal.

4. Any evidence that is entered in the evidence system and is not used in judicial or administrative action may be disposed of after a period of six months, or sooner if it becomes obvious that it has no evidentiary value whatsoever or if early disposal is directed by the CO or SJA. Extreme care must be taken in early disposal since the original incident may later indicate that the evidence should have been retained.

5. In significant unresolved cases evidence should be retained until expiration of the statute of limitations. The SO may authorize such disposal after consulting with the requestor and the SJA. In such cases, the evidence custodian shall complete the final disposition section of the ECD showing the SO as the authorizing official.

6. Active evidence belonging to ships, stations, or units scheduled for decommissioning or closure shall be transferred to the nearest active evidence locker, as specified by the area commander/coordinator, immediate superior in command (ISIC), or Echelon 2 commander. Prior to the transfer, all evidence that can be reasonably disposed of shall be cleared from the locker and appropriate ECD entries made.

I.9 EVIDENCE DISPOSAL GUIDELINES

1. Evidence that is the personal property of an individual shall, whenever possible, be returned to that person, with the exception of contraband or other unlawful items. When personal property is returned to the owner

or his/her authorized representative, the individual receiving the property shall be required to sign in the disposition section of the original ECD (or copy, if the original is absent). If the owner or representative presents a property receipt when making the claim, the receipt will be obtained and destroyed. In the event that the owner refuses to accept all the property seized, this will be noted on the ECD and other appropriate disposal will be made of the property. In the event that certain personal property the possession of which is not generally unlawful but is prohibited by command or installation orders is entered into the evidence custody system, when it has served its purpose the property shall be returned to the command having control over the individual from whom it was obtained. It will be necessary for that command to provide a receipt for the property and make a determination as to its disposition.

2. When evidence has been received that is the custodial responsibility of the authority requesting the investigation, it shall be returned to the requester's representative, and that person shall be required to acknowledge receipt for it in the final disposition section of the ECD.

3. All U.S. Government property that cannot be identified as belonging to a particular activity or command shall be submitted to the activity supply department or the nearest Navy supply activity. In addition to any documentation required by the receiving activity, the activity's representative shall acknowledge receipt for the material in the final disposition section of the ECD. In the event that the activity declines to provide a receipt for the property on the ECD, a suitable receipting document will be obtained and attached to the ECD.

4. U.S. Government currency/negotiable instruments that cannot be returned to a rightful owner shall be turned in to the U.S. Treasury via the activity disbursing officer.

5. Final disposition on government-owned weapons will be IAW and comply with OPNAVINST 5530.13 (series), Department of the Navy Physical Security Instruction for Conventional Arms, Ammunition, and Explosives (AA&E).

6. Evidence that cannot be returned to the owner and has been entered into the Navy evidence custody system for disposal, such as drugs, illegal firearms, or other contraband, shall be destroyed. Such destruction shall be accomplished by or in the presence of the evidence custodian or alternate custodian and a disinterested third party, both of whom shall sign the final disposition section of the ECD.

7. Under no circumstance will any evidence be converted for use by a Navy component or for the personal use of any individual within or without DON. Releasing controlled-substance evidence for use as training aids to Navy drug detector dog teams or for the purpose of controlled burns or making training aid display boards is not authorized.

INTENTIONALLY BLANK

APPENDIX J

Investigations

J.1 GENERAL

Criminal investigations are official inquiries into crimes involving the military community. A criminal investigation is the process of searching, collecting, preparing, identifying, and presenting evidence to prove the truth or falsity of an issue of law. Investigators conduct systematic and impartial investigations to uncover the truth. They seek to determine whether a crime was committed and to discover evidence of who committed it. Investigators' efforts are focused on finding, protecting, collecting, and preserving evidence discovered at the crime scene or elsewhere. For criminal investigations to be successful, the investigator must understand the general rules of evidence, provisions and restrictions of the most current MCM, and the UCMJ. Training requirements for command criminal investigator (CCI) personnel are identified in OPNAVINST 5530.14 (series), Navy Physical Security and Law Enforcement Program.

J.2 ROLE AND RESPONSIBILITIES

1. NCIS is responsible for facilitating and establishing policy for the Navy's Law Enforcement Investigative Program. The primary goal of this program is to ensure the Navy LE operates with a single investigative standard.

2. NCIS may assign supervisory special agents as Regional Investigations Coordinators (RICs) by region, with a primary mission focus of providing oversight to the investigative program. RIC oversight includes CCI screening, selection, and investigative standardization.

3. CCIs are primarily tasked to investigate all crimes, offenses, or incidents falling within their investigative jurisdiction, including criminal investigations involving UCMJ violations and other criminal acts that are not pursued by NCIS. CCI refers to both military and civilian personnel.

J.3 OBJECTIVES OF LAW ENFORCEMENT INVESTIGATIONS

Objectives of criminal investigations are as follows:

1. Determine whether a crime was committed.

2. Collect information and evidence legally to identify who was responsible.

3. Apprehend the person(s) responsible or report them to the appropriate civilian police agency (where applicable).

4. Recover stolen property.

5. Present the best possible case to the prosecutor.

6. Provide clear, concise testimony.

J.4 JURISDICTION

1. The CCI receives the authority to conduct criminal investigations pursuant to the UCMJ and other criminal acts from the CO.

2. CCIs are limited to the same jurisdictional authority granted other NSF members.

3. Work with NCIS special agents does not convey special agent authority and jurisdiction to a CCI.

4. During the conduct of some investigations, the CCI may be required to perform official duties off base and in the civilian community.

J.5 CREDENTIALS

1. The SO will issue CCI credentials that are signed by the CO. CCI credentials (OPNAV 5580/26) shall be issued upon assumption of duties, and bearers shall maintain these credentials during the tenure of their CCI assignment. SO and RIC personnel will work in concert to ensure successful completion of the screening/selection and training process by the CCI prior to the issuance of credentials. Credentials shall be serially numbered and contain the name, a color photograph, and a physical description (i.e., height, weight, eye and hair color, gender, and date of birth). The CO will sign the credentials, thus duly appointing the bearer as a command investigator. Upon incurring a permanent change of station transfer, termination, or reassignment, the CCI shall return the credentials.

2. Credentials shall be strictly controlled by serial number. Logs shall fully identify the person to whom issued and shall record the disposal/destruction action of expired or returned credentials. Blank credentials shall be securely stored and shall be inventoried annually or during SO/investigator supervisor turnover inventories.

3. The CCIs are responsible for safeguarding their issued credentials. If credentials are misplaced or stolen, the CCI must report this loss immediately to his/her supervisor. The SO and RIC shall be made aware of all lost or stolen investigator credentials.

J.6 NAVAL CRIMINAL INVESTIGATIVE SERVICE REFERRALS/DECLINATIONS

The SO should ensure that prompt referrals of investigations are made to NCIS and that regular lines of communications are maintained. Upon an NCIS declination, the command criminal investigative division (CCID) or patrol section will assume the case for investigative action. Security department roles, responsibilities, and procedures will provide the standard by which patrol rather than CCID is used to investigate such events. Where appropriate, the RIC will be consulted with regards to investigative procedures and strategy. In a remote environment, the CCI will assume the investigation and consult with the nearest NCIS special agent or area RIC.

J.7 ADJUDICATIVE NOTIFICATION

1. Briefing appropriate command personnel regarding the investigative results of a case is an important phase of any investigation.

2. Upon completion of an investigation, periodic updates/status reports or, as circumstances dictate, the CCID supervisor or investigator will offer a comprehensive brief to the command that maintains convening authority over the suspect. The adjudicative notification should be timely, concise, and factual. The CCI should articulate the facts of the case as they developed during the course of investigative activity.

3. The final investigative report can be provided to the command during this briefing. This process does not negate the need to occasionally provide investigative summary briefs to other interested parties with the concurrence of the respective commander.

4. Every effort should be made to effect contact with SJAs, trial attorneys, or other commands that have personnel directly involved in the investigation.

J.8 INVESTIGATIVE PROTOCOLS

1. General. NCIS investigative procedures, protocols, and guidelines are the SOPs for all Navy CCI personnel and units. Through a robust in-service training program and continuous liaison with area NCIS field offices, the RIC shall ensure that command investigators are kept apprised of current investigative methods and procedures.

2. Interviews and interrogations. The interview of a victim or witness and the interrogation of a suspect are intricate parts of the investigative process. Interviews and interrogations provide the investigator with a unique opportunity to ascertain the facts of a case and successfully resolve the investigation.

   a. An interview and interrogation log (OPNAV 5580/7) shall be completed prior to any session. Interviews and interrogations should be thoroughly planned out, and consultation with RIC personnel regarding specific methods and/or strategies is recommended.

   b. A signed, sworn statement shall be taken after any interview or interrogation wherein pertinent information is generated. If for any reason the subject refuses to sign the statement, it will be included in the investigative report as the CCI's summary of interview/interrogation.

3. Narcotics operations. CCIs shall not initiate or engage in any operational activity involving the sale, transfer, and/or distribution of controlled substances without the prior approval of the commander and the area RIC or NCIS special agent in charge. CCIs are encouraged to assist area NCIS offices in these operations in a support capacity. This operational prohibition does not apply to routine narcotics investigations.

4. Theft-suppression operations. Theft-suppression operations are considered covert surveillance activities that are normally manpower-intensive. These operations are used when high-volume theft activity is identified within a certain area. Several operational strategies and tactics exist, and the RIC should be consulted prior to initiation so that the best possible scenario can be used. These special operations can be performed aboard a ship or shore station. The SO and RIC will approve any theft-suppression operation.

5. Cooperating witnesses. The use of cooperating witnesses (CWs) can be beneficial in successfully resolving an investigation. However, the use of an NCIS-registered CW during the conduct of a CCI investigation will be considered only on a case-by-case basis and with prior consultation of the RIC or area NCIS field office.

   a. CCIs shall not engage in the registration, use, or handling of a CW. During the conduct of an NCIS-generated investigation wherein a CW is being used and a CCI is participating, careful attention must be made to maintain CW safety and operational security.

   b. The CCI shall not brief any aspect of a CW's use outside the confines of NCIS. NCIS personnel shall perform command briefings involving CW activities.

6. Records checks. Conducting records checks and criminal history queries are standard investigative protocols and are encouraged to be used by CCI.

   a. Potential databases that can be queried include NCIC, the Defense Central Index of Investigations (DCII), and NLETS. Only those organizations and components that maintain connectivity and are registered within the respective database can request and retrieve information. These databases are used solely for LE purposes.

   b. Other applications, such as background checks, employment verification, and criminal history checks for employment, are routinely prohibited. NSF without NCIC/NLETS access can contact their servicing NCIS office for query assistance.

J.9 NAVAL CRIMINAL INVESTIGATIVE SERVICE TECHNICAL SUPPORT

J.9.1 General

Electronic recordings, video surveillance, mail covers/pen registers, and polygraph use are considered advanced investigative techniques and shall be pursued only with the assistance and direction of NCIS. The CCI shall notify the SO and RIC of all activities involving these investigative techniques. Special considerations and reporting requirements exist for these activities.

J.9.2 Electronic Communications Intercepts

CCI shall not engage in the consensual or nonconsensual recordings/intercepts of one or more person's conversations by electronic means. NCIS is designated as the only DON component authorized to conduct oral and/or wire intercepts for LE purposes. This prohibition does not apply to consensual video recording of routine interviews/interrogations during investigative activity.

J.9.3 Video Surveillance

Video surveillance is the means by which the activities and actions of individuals are covertly recorded via a photographic device. Video cameras are normally hidden in concealed locations for the purpose of capturing and recording criminal activity. The use of video surveillance equipment during the course of operational activity is encouraged. The SO must authorize the use of covert video surveillance activity/equipment. The RIC shall be consulted prior to the initiation of covert video surveillance. Although rules governing video surveillance are not as restrictive as other electronic recording techniques, some legal and administrative requirements exist and must be met prior to the initiation of operational activity. Permanently mounted CCTV equipment within an installation is not considered covert video surveillance for the purposes of this section. The consensual photographic recording of a routine interview/interrogation is not considered video surveillance.

J.9.4 Mail Covers/Pen Registers

A mail cover is an investigative process in which certain information and data are obtained from mail parcels and letters. A pen register is a device that records electronic impulses that identify certain numbers transmitted on a telephone line. The SO and the RIC shall be made aware of all activities evolving mail covers and pen registers. Various legal, administrative, and reporting requirements must be met prior to use. The RIC or area NCIS office shall be directly involved in any cases evolving these investigative techniques.

J.9.5 Polygraph Examinations

The use of polygraph examinations is encouraged during investigative activity. NCIS maintains numerous polygraph offices that are accessible to the CCI. CCIs will request polygraph support through the RIC or area NCIS office. Specific legal and administrative guidelines must be met prior to use.

J.10 NAVAL CRIMINAL INVESTIGATIVE SERVICE FORENSIC SUPPORT

The CCI should use every resource available in resolving criminal investigations. NCIS has access to forensic laboratories that provide narcotic, fingerprint, trace evidence, and questioned-document examinations and analysis. Forensic laboratory use is encouraged and available to the CCI. The RIC will provide assistance and guidance with regard to the proper laboratory use.

J.11 FINGERPRINTING SUSPECTS

Criminal fingerprint cards (FD-249) will be submitted for all military Service members investigated for the commission of any offense IAW DODI 5505.11, Fingerprint Card and Final Disposition Report Submission Requirements, when a command initiates military judicial proceedings or when command action is taken in a non-judicial punishment proceeding. The use of live scan or other electronic means to obtain and submit fingerprints is authorized when available.

J.12 INVESTIGATIVE REPORTING

1. The IR is drafted under the provisions of CLEOC reporting format and shall contain relevant statements, investigative action summaries, and other documentation pertinent to the case.

   a. Report of investigation (ROI) shall be prepared unless special circumstances dictate otherwise at the completion of each investigation. The ROI, a comprehensive summary of all investigative actions, containing original signed, sworn statements and other original documents collected during the course of investigative activity, shall be provided to the prosecuting command.

   b. Signed, sworn statements, in their original state, will be needed for adjudicative purposes and are therefore provided to the prosecuting command. Copies of the ROI may be disseminated to other interested commands. Pertinent SJAs should be consulted prior to distribution outside of the command.

2. Statements contained within the ROI should include victim/witness interviews and/or suspect interrogations. Documents such as reports, financial records, logs, journals, or any other document of perceived importance should be included in the ROI. The RIC will assist the CCI in the compilation of any ROI or investigative summary.

3. Case management. The CCI shall maintain a separate case file for every investigation that is initiated. The RIC will provide guidance and instruction on the composition and maintenance of case files.

   a. The case file should include all patrol and investigative reports, interview/interrogation notes and logs, statements, official records and documents, and any other material deemed pertinent to the case. The material contained within the case file should be situated in such a way as to allow for a clear, coherent, and logical review process. The case file shall include a summary sheet that documents each case review session and notes any inputs or recommendations the investigator supervisor/RIC elects to make.

   b. All case files are to be centrally maintained and secured within the criminal investigation division (CID). Retention and disposition of case files are pursuant to SECNAVINST 5212.5 (series), Navy and Marine Corps Records Distribution Manual. The RIC will conduct an annual case management inspection to ensure recordkeeping compliance.

4. Case review/quality assurance. The RIC shall ensure that an effective program is in place to ensure the quality assurance review and maintenance of investigative case files. These case reviews are important for the timely and logical flow of investigative activity. Case reviews are used to gauge the quality of investigative efforts and provide input relative to the successful resolution of an investigation. The supervisor within the CID should conduct these reviews in person and on a biweekly basis, or when other circumstances mandate a more expedient review. The RIC will engage in periodic case reviews with command investigators in an oversight capacity. The results of all case/quality assurance reviews shall be documented and maintained within the case file. Significant discrepancies shall be discussed with the SO.

J.13 SAFETY CONSIDERATIONS FOR INVESTIGATORS

Personal safety is a priority for investigators. Responding to a crime scene often places the investigator in danger of exposure to hazards or unsafe conditions. Personal protection against biohazards, chemical hazards, and physical hazards requires special training and equipment. Law enforcement investigators should receive HAZMAT and critical incident management training.

J.14 VICTIM AND WITNESS ASSISTANCE

1. LE investigators are required to inform victims and witnesses of the services available to them. Particular attention should be paid to victims of serious and violent crimes, including child abuse, domestic violence, and sexual misconduct.

2. DODD 1030.01, Victim and Witness Assistance, and DODI 1030.2, Victim and Witness Assistance Procedures, implement statutory requirements for victim and witness assistance and provide guidance for assisting victims and witnesses of crime, from initial contact with offenders through investigation, prosecution, and confinement. Together, the directive and instruction provide policy guidance and specific procedures for all sectors of the military to follow.

   a. The directive includes a bill of rights that closely resembles the federal crime victims' bill of rights. Investigators, as well as all DOD officials, are responsible for ensuring that victims of military crimes have these rights. Victims are to be:

      (1) Treated with fairness and respect

      (2) Reasonably protected from the offender

      (3) Notified of court-martial proceedings

      (4) Present at court-martial proceedings

      (5) Able to confer with the government attorney

      (6) Given available restitution

      (7) Informed of the outcome of an offender's trial and release from confinement.

   b. VWAPs cover the entire military justice process, from investigation through prosecution and confinement. Investigators must ensure proper documentation as prescribed in the DOD guidance throughout all stages of the case.

J.15 LEGAL PROTECTION OF JUVENILES

1. A juvenile is identified as a person who is less than 18 years of age at the time of the incident being investigated and who is not active duty military. Certain precautions must be observed when questioning juveniles, regardless of whether they are the victim, witness, suspect, or subject. Investigative steps for the gathering of evidence in juvenile offenses are the same as those used in cases involving adult suspects. Ensure the juvenile is processed according to Chapters 401 and 403, Part IV, Title 18, U.S.C.

2. A civilian Miranda warning must be given to a juvenile in terms that he or she can understand. The warning should be given in the presence of a parent, guardian, or advocate IAW applicable laws.

3. Detained juvenile suspects cannot be placed in confinement facilities or detention cells. Juveniles may be temporarily detained in police administration office areas or those areas authorized by the SJA. Unless a juvenile is taken into custody for serious offenses, do not take any fingerprints or photographs without written parental or judicial consent. Do not release any names or pictures of juvenile offenders to the public.

4. Records of juvenile offenders must be secured and released only on a need-to-know basis. During juvenile proceedings, data on the juvenile and the offense may be given only to the court, the juvenile's counsel, and others having a need to know. Others may include courts or agencies preparing presentence reports for other courts, or they may be police agencies requesting the information for the investigation of a crime.

5. Permanent records are not made for nonessential minor incidents or situations resolved in conference with parents of the juvenile. If a juvenile is found innocent, all records of the offense (including fingerprints) must be destroyed, sealed by the court, or disposed of according to directives.

J.16 INVESTIGATIVE PROTOCOLS

Conducting a successful investigation is often the result of having a wide range of knowledge and using common sense in its application. There are certain actions that apply to all investigations. The legal requirements concerning evidence are met when the investigator can identify each piece of evidence, describe the exact location of the item, indicate when the item was collected, and maintain and show a proper chain of custody. An investigator's notes, photographs, and sketches become the legal record of his actions. The written and photographic records of the scene also put the presented evidence into perspective and help laboratory experts determine what tests are needed. Finally, the investigator's notes, photographs, and sketches provide him with a valuable reference for his actions during the processing of the crime scene.

J.17 SURVEILLANCE DETECTION OPERATIONS

The increase in reporting of suspicious individuals conducting surveillance of DOD sites in the United States and overseas indicates possible preoperational targeting by terrorists and merits attention by commanders at all levels. Commanders and security planners need to understand how terrorist surveillance is conducted and the purpose of terrorist surveillance operations. With this knowledge, commanders can implement protective countermeasures; comply with DOD standardized reporting procedures; and deter, detect, disrupt, and defend against future attacks. Surveillance is the deliberate, systematic, and continuous process of monitoring a person, place, or thing. Surveillance can be conducted overtly or covertly. Surveillance detection (SD) is the deliberate, systematic, and continuous process of identifying the presence or absence of surveillance through observation by applying the time-distance-direction rule.

J.17.1 Effective Surveillance Detection Program Elements

An effective SD program is built on a foundation of accumulated information, threat analysis, and preparation. Commanders are responsible for implementing a well-designed AT plan that is capable of detection, assessment, delay, command and control, and response.

J.17.2 Responsibilities

Commanders have an inherent responsibility for protection planning, operations, and training within their commands. Included within protection planning is the development of an SD program for regions and installations. It is the responsibility and duty of all commanders to protect DOD personnel, facilities, and material under their control from terrorist attacks and to mitigate the effects of an attack should one occur. When developing an SD program, commanders should ensure procedures are established to report up and down the chain of command and through the regional NCIS office all information pertaining to suspected surveillance.

J.17.3 Terrorist Surveillance Techniques

Reporting indicates that terrorists normally carry out detailed preoperational planning, including surveillance, prior to conducting an act of terrorism. There are seven distinct steps in the terrorist attack cycle:

1. Step 1: Target Selection. This phase initiates the operational cycle timeline. During this phase potential targets are chosen.

2. Step 2: Surveillance. Potential targets are placed under surveillance to determine the final target.

3. Step 3: Final Selection. Surveillance assessment data are evaluated and analyzed to identify the target. This is a key point in the operational cycle timeline, during which all data are assimilated and a specific target is selected.

4. Step 4: Planning. Specifics of the attack are determined. In this step the target is clearly defined. The target and the surrounding area are now the object of a detailed reconnaissance to determine vulnerabilities (gaps and seams) and to identify the target's security measures, defenses, and potential obstacles that might hinder ingress and egress.

5. Step 5: Final Surveillance. Prior to deployment of the terrorist attack element, surveillance is conducted to verify information collected earlier and to familiarize the terrorists with the attack plan. It is during this step that final surveillance of the target may occur, to determine whether any last-minute security procedures have been put into place (Jersey barriers, police, etc).

6. Step 6: Deployment. If there is no change to the information, the terrorist attack element will deploy to the selected attack site for execution of the plan. If there are changes, terrorists will be forced to abandon or amend their plan.

7. Step 7: Attack. The type of attack (close in or standoff) as well as the attack site and timing is predicated on the information gathered by the terrorists in steps 2 and 5 and must offer plausible opportunity for success.

J.17.4 Forms of Physical Surveillance

There are three basic forms of physical surveillance—foot, vehicle, and stationary—generally categorized as mobile or static. Commands should be more concerned with static (fixed) surveillance.

1. Mobile surveillance involves following a targeted individual either on foot or by vehicle and usually progresses in phases from a stakeout to a pickup point and then through the follow phase until the target stops. At this point operatives are positioned to cover logical routes, enabling surveillance to continue when the target moves again.

2. Static, or fixed, surveillance involves conducting surveillance from one location to observe a target, whether a person, building, facility, or installation. Fixed surveillance often requires the use of an observation point (OP) to maintain constant, discreet observation of a specific location. OPs can be established in houses, apartments, offices, or stores, or on the street. A mobile surveillance unit, such as a parked car or van, can also serve as an OP. Terrorists may park a vehicle outside a building, facility, or installation to observe routines of security and personnel. Terrorists may move on foot or use various means of transportation, including buses, trains, and boats.

J.17.5 Surveillance Detection Principles

Because terrorists usually conduct surveillance (often over a period of weeks, months, or years), detection of their activities is possible, regardless of the level of expertise of the terrorist conducting surveillance. Knowing what to look for and distinguishing the ordinary from the extraordinary are keys to successful SD. For these reasons, SD, in its most basic form, is simply watching for persons observing personnel, facilities, and installations.

The objectives of SD are to record the activities of persons or vehicles behaving in a suspicious manner and to provide this information in a format usable by the appropriate law enforcement or intelligence officials (normally NCIS). Personnel using SD techniques should avoid confrontations with individuals displaying suspicious behavior, except in exigent circumstances or in situations described in paragraph J.17.9. Depending on increased indicators and warnings, the installation's Threat and Antiterrorism Threat Working Groups may determine the need for specialized covert SD measures to ensure installation protection. If this occurs, the command shall consult with regions, installation commanders, NCIS, and local LE officials to develop an appropriate program.

For SD efforts to achieve positive results, NSF personnel should immediately report all incidents of possible surveillance and suspicious activities. This reporting shall provide detailed descriptions of the people, time of day, locations, vehicles involved, and circumstances of the sightings, through the chain of command to the local NCIS office. This information may then be incorporated into a NCIS suspicious incident report. These reports are important pieces of information that, when combined with other similar sightings, allow security personnel to assess the level of threat against a specific facility, installation, or geographic region.

The emphasis of SD is on indicators and warnings of terrorist surveillance activities. SD efforts will focus on overtly recording and then reporting incidents similar to the following:

1. Multiple sightings of the same suspicious person, vehicle, or activity, separated by time, distance, or direction.

Note

Time-Distance-Direction Rule. If—over long enough periods of time, a far enough distance, and a nonlinear enough area of travel—a vehicle or person is noticed on at least three separate occasions, then surveillance is most likely being conducted.

2. Suspicious activity at a location previously designated as a possible terrorist surveillance OP through a detailed target analysis.

Note

Target analysis and vulnerability assessment matrixes should be consistent with the principles and methodologies outlined in DOD O-2000.12H, Department of Defense Antiterrorism Handbook (FOUO).

3. Individuals who stay at bus stops for extended periods while buses come and go.

4. Individuals who carry on long conversations on pay or cellular telephones.

5. Joggers who stand and stretch for an inordinate amount of time.

6. Individuals sitting in a parked car for an extended period of time.

7. Individuals who do not fit into the surrounding environment by wearing improper attire for the location or season.

8. Individuals drawing pictures, taking notes in an area not normally of interest to a standard tourist, showing interest in or photographing security cameras or guard locations, or noticeably watching security reaction drills and procedures.

9. Individuals who exhibit unusual behavior, such as staring or quickly looking away from individuals or vehicles as they enter or leave designated facilities or parking areas.

10. Terrorists may also employ aggressive surveillance by false telephone threats, approaching security checkpoints to ask for directions, or innocently attempting to smuggle nonlethal contraband through checkpoints. These activities are conducted to test the effectiveness of search procedures and to gauge the alertness and reaction of security personnel.

It is important to highlight that the above surveillance indicators are to be observed and recorded overtly while performing normal security force activities. The intent is to raise the awareness of our security forces to record and report suspicious incidents during the course of routine LE and security duties.

J.17.6 Surveillance Detection Training

1. The following are minimum training requirements for NSF or contractor personnel assigned to conduct SD:

   a. Graduate of DON Police Academy or CNIC Regional Training Academy using the Navy Security Force Training Course curriculum or MAA "A" School

   b. Graduate of DON-approved SD program of instruction (normally given by NCIS)

   c. Receive sufficient training by command Judge Advocate General on military/civilian interactions; Federal Tort Claims Act (28 U.S.C. 1346); Posse Comitatus Act (18 U.S.C. 1385); DODD 5200.27, Acquisition of Information Concerning Persons and Organizations Not Affiliated with the Department of Defense; jurisdiction of the installation; federal, state, and local criminal laws; and other legal issues and laws as designated by the command Judge Advocate General.

2. Additional requirements are as follows:

   a. Be able to work independently with little supervision

   b. Be knowledgeable of city and state criminal laws as they pertain to DOD and civilian interactions.

J.17.7 Surveillance Detection Reporting

Implementing effective security and employing SD principles should deter terrorist surveillance. However, regardless of the capabilities of a facility or installation to provide protective measures, good working relationships with local, state, and federal law enforcement agencies are essential in establishing cohesive, timely, and effective responses to indications of terrorist activity. Regional commanders, installation commanders, and senior LE officials should coordinate with NCIS to establish partnerships with local civilian agencies to develop intelligence- and information-sharing relationships. If indicators of terrorist surveillance exist, the SD program should provide detailed reports of the indicators to the appropriate LE or intelligence activity (normally NCIS). As reports of suspicious activity increase and trends are interpreted by intelligence or LE personnel to indicate preoperational terrorist surveillance, it may be necessary to coordinate with NCIS and other LE agencies to implement a more sophisticated SD program. Specialized SD assets should be coordinated through the appropriate chain of command to NCIS.

Ensure that SD information is reported up the chain of command. Region and installation commanders shall implement the following:

1. Security personnel assigned to conduct SD shall fill out the SD log for each shift and submit all surveillance detection information cards (SDICs) as necessary. See figure J-1 for an example of the SDIC.

2. All SDICs shall be evaluated by the installation chief investigator and submitted to the installation's security officer when necessary. Additionally, information should also be forwarded to installation commanders for review and to the regional and installation AT office for dissemination when appropriate.

3. If any suspected surveillance is noted, a report shall be transmitted to the local NCIS office for review, with a copy sent to the appropriate combatant command (Attn: N34).

NCIS will determine what information, if any, to share with the local authorities via designated format and will possibly generate a suspicious incident report. If deemed necessary by NCIS, the suspicious incident report may be entered as a GUARDIAN Report (formally known as TALON) and transmitted to the NCIS Multiple Threat Alert Center via the local FBI Joint Terrorism Task Force.

181

SURVEILLANCE DETECTION INFORMATION CARD

Date	Time	Location

Suspicious Person:

Sex	Race	Height	Weight	Eye	Nationality	Age	Scars, Marks, Tattoos

HAIR
- Color
- Length
- Type
- ☐ Beard
- ☐ Sideburns
- ☐ Mustache

VOICE
Language:
- ☐ Loud
- ☐ High-Pitch
- ☐ Soft
- ☐ Deep
- ☐ Fast
- ☐ Stutter
- ☐ Slow
- ☐ Slurred
- ☐ Nasal

Accent:

ITEMS
List (i.e., backpack, camera etc.)
_____
_____
_____
_____
_____

CLOTHING

UPPER — Type / Color/Description
- ☐ Coat _____
- ☐ Vest _____
- ☐ Shirt _____
- ☐ Tie _____
- ☐ ___ _____

LOWER — Type / Color/Description
- ☐ Pants _____
- ☐ Shoes _____
- ☐ Belt _____
- ☐ ___ _____
- ☐ ___ _____

OTHER — Type / Color/Description
- ☐ Eyeglasses _____
- ☐ Hat _____
- ☐ Gloves _____
- ☐ ___ _____
- ☐ ___ _____

Suspicious Vehicle:

Color	Make	Model	Body Style	License #	State

Other identifying information (i.e., damage, tinted windows, clean, dirty, etc.)
_____
_____

Specific Actions/Behavior displayed by person/vehicle:

_____
_____
_____
_____
_____
_____

Person who observed Surveillance:

Name (Last, First MI)	Command	Contact #

Figure J-1. Surveillance Detection Information Card

J.17.8 Surveillance Detection Procedures

1. The SD program will follow the general guidelines listed below:

    a. COs and security departments shall coordinate the training and implementation of any SD operation with the regional NCIS office and installation and regional staff judge advocate prior to the initiation of a SD mission. SD operational planning should be executed, designed, and briefed using the standard five-paragraph order format.

    b. Security or contractor personnel assigned to conduct SD shall have completed training required in paragraph J.17.6 of this instruction.

    c. Security personnel shall be fit for full security duty. No limited duty or injured-in-line-of-duty personnel will be used.

    d. Privately owned vehicles shall not be used during SD operations.

2. Unless the need for covert SD measures is approved by the CO, NCIS, and local LE, personnel will not set up SD points off government property. However, personnel must go off the installation to effectively conduct a target analysis of the installation that will identify possible terrorist surveillance OPs. Prior coordination between the security officer and local authorities is necessary during this phase of planning.

3. At a minimum, SD personnel shall have the following equipment:

    a. Two-way radio

    b. Binoculars

    c. Phone list with NCIS and local police

    d. Camera (optional)

    e. Surveillance detection information cards.

J.17.9 Responding to Photographic Surveillance

1. Post 9/11, security departments have been particularly concerned with individuals photographing Navy installations and vessels from locations outside DOD jurisdiction (off base), either aboard private/commercial watercraft or ashore. After inquiry, most incidents have been civilians participating in sightseeing activities, although some incidents may have been personnel engaged in surveillance. As a result of the concerns that these incidents have caused among Navy security departments, specific guidance on responding to these situations is provided.

2. When determining what actions are permitted when responding to suspicious activity off-base, one must consider the Posse Comitatus Act (18 U.S.C. Section 1385) and applicable DOD Directives. The PCA generally prohibits federal military personnel from acting in a law enforcement capacity, exercising nominally state law enforcement, police, or peace officer powers that maintain law and order on nonfederal property, except where expressly authorized by the U.S. Constitution or Congress. However, the U.S. Congress has enacted a number of exceptions to the PCA that allow the military, in certain situations, to assist civilian law enforcement agencies in enforcing the laws of the United States. The most common example that may apply to suspicious off-base photographic surveillance is known as the military purpose doctrine (MPD). The MPD is an exception to the PCA that applies to actions that are taken for the primary purpose of furthering a military or foreign affairs function of the United States. As identified in DODD 5525.5, this doctrine includes protection of DOD personnel, DOD equipment, and official guests of DOD and may allow military personnel to engage suspicious persons off-base for the purpose of force protection.

3. If someone is seen photographing Navy installations and/or vessels off-base outside DOD jurisdiction ashore, security personnel shall immediately report such behavior to the chain of command, NCIS, and, if necessary, appropriate civilian law enforcement personnel. Information gathered shall comply with DODD 5200.27, Acquisition of Information Concerning Persons and Organizations Not Affiliated with the Department of Defense, which limits the "acquisition of information on individuals or organizations not affiliated with the DOD . . . to that which is essential to the accomplishment of assigned DOD missions" and "no information shall be acquired about a person or organization solely because of lawful advocacy of measures in opposition to Government policy." Professionalism and courtesy are necessary when responding to and speaking with an individual who is taking pictures. The primary goal of this contact is to assess if the individual(s) has a legitimate purpose for the photography, such as journalism or sightseeing. Regardless of the justification provided by the photographer, information on the event should be forwarded to NCIS, since what appears to be innocent picture-taking may be part of a greater pattern of which the local security staff is unaware.

4. If someone is seen photographing Navy installations and vessels off-base outside DOD jurisdiction ashore, a marked security patrol unit should be dispatched to approach the individual(s) and inquire why they are taking pictures of DOD assets. Security personnel should ask the individual(s) for identification and advise them of the Navy's concerns regarding security. Throughout the exchange, security personnel must pay careful attention to the behavior of the individual(s). If suspicious behavior is noted, security personnel shall immediately report such behavior to the chain of command, NCIS, and, if necessary, appropriate civilian LE. Individuals are not to be apprehended but may be asked to remain where they are until civilian LE arrives. The seizure of photographic film or equipment is not authorized; however, security personnel may request the film/equipment pending a determination of the nature of the threat, if any. If no suspicious behavior is noted, the photographer(s) should be thanked for understanding and cooperating and permitted to depart. The involved security personnel should then report any such off-base questioning of civilians to their chain of command.

5. Installation commanders may promulgate instructions to prohibit on-base photography when such prohibitions are reasonably necessary to protect genuine security interests. Prohibitions should be limited to areas that contain especially sensitive facilities (communication nodes, antenna farms, weapons-related storage or handling areas) or access points (gates and guard positions). Broad prohibitions of all photography on an installation are generally not necessary and should be avoided. Areas in which photography is prohibited must be properly posted to notify members of the public and military personnel of the prohibition.

6. On DOD property, any persons observed taking photographs or any other action that may constitute surveillance activity may be approached by security personnel and questioned to determine intent. This principle applies regardless of whether the area involved has been posted as a "No Photography" area.

INTENTIONALLY BLANK

APPENDIX K

Nonlethal Weapon Procedures

K.1 NONLETHAL WEAPONS

NSF personnel, although always armed when performing LE duties, may also be armed with NLW or equipment. Even though their intended purpose is nonlethal, these weapons could cause death or serious bodily harm. All NSF shall be trained and equipped with a minimum of hand restraints, baton, and OC. For further guidance, refer to FM 3-22.40/MCWP 3-15.8/NTTP 3-07.3.2/AFTTP(1) 3-2.45, Multiservice Tactics, Techniques, and Procedures for the Tactical Employment of Nonlethal Weapons.

K.2 POLICY

DODD 3000.3, Policy for Non-lethal Weapons, states, …nonlethal weapons, doctrine, and concepts of operation shall be designed to reinforce deterrence and expand the range of options available to commanders. Non-lethal weapons should enhance the capability of U.S. forces to accomplish the following objectives:

1. Discourage, delay, or prevent hostile actions.

2. Limit escalation.

3. Take military action in situations where use of lethal force is not the preferred option.

4. Better protect our forces.

5. Temporarily disable equipment, facilities, and personnel.

6. Nonlethal weapons should also be designed to help decrease the post conflict costs of reconstruction.

DODD 3000.3 continues by stating, The availability of nonlethal weapons shall not limit a commander's inherent authority and obligation to use all necessary means available and to take all appropriate action in self-defense.

Neither the presence nor the potential effect of nonlethal weapons shall constitute an obligation for their employment or a higher standard for employment of force than provided for by applicable law. In all cases, the United States retains the option for immediate use of lethal weapons, when appropriate, consistent with international law.

The directive further states, Nonlethal weapons shall not be required to have a zero probability of producing fatalities or permanent injuries. However, while complete avoidance of these effects is not guaranteed or expected, when properly employed, nonlethal weapons should significantly reduce them as compared with physically destroying the same target.

Nonlethal weapons may be used in conjunction with lethal weapon systems to enhance the latter's effectiveness and efficiency in military operations. This will apply across the range of military operations to include those situations where overwhelming force is employed.

K.3 EMPLOYMENT

The following are relationships of NLW and nonlethal capabilities, up to the use of deadly force.

1. The possession of NLW shall not prevent NSF personnel from using deadly force when needed. Neither the presence nor the potential effect of NLW shall constitute an obligation for their employment or a higher standard for employment of force than provided in Appendix B of this NTTP.

2. The use of NLW shall not be required to have a zero probability of producing fatalities or permanent injuries. However, while complete avoidance of these effects is not guaranteed or expected, when properly employed, NLW should significantly reduce them. The following provisions apply:

   a. Handcuffs. Handcuffs shall be securely fastened but not so tightly as to cause the individual injury. When in use, handcuffs will be double-locked and checked periodically to ensure they are not causing injuries. Suspects shall not be handcuffed to vehicles. Handcuffing to other objects, such as interrogation desks, should be considered only when suspects are deemed to be a danger to themselves or the interviewer. These precautions are also applicable to the use of flex cuffs and leg irons.

   b. Batons. NSF personnel shall be trained in the target areas of the human body. Issuance of a baton will be authorized only after formal training. Expandable batons are the preferred NLW.

K.4 ESCALATION-OF-TRAUMA CHART

K.4.1 Description

Figure K-1 is a representation of the human body using white, light gray, and dark gray areas depicting the less likely to most likely injury prone areas.

K.4.2 Guidance

Figure K-1 fosters a learned capacity to recognize levels of possible trauma, from white (lowest level) to light gray (moderate level) to dark gray (highest level), which benefits an NSF's performance in training and in the field. Guidance is what the figure is all about. It incorporates complex biomedical considerations and combines them into a humane, sensible, versatile, easy-to-learn, easy-to-recall way to use blunt impact in a confrontation.

K.4.3 Introduction

The program manager of Monadnock Lifetime Products, Inc., authorized the use of the company's copyrighted material in figure K-1 on the condition that it include the following associated background and concept in action paragraphs, in addition to a few general-policy statements applicable to the Monadnock Police Training Council. (See paragraphs K.4.4. and K.4.5.)

1. The use of force by officers is permissible when used to effect an apprehension; to overcome resistance; to prevent escape; in self-defense; or in the defense of others. The force used must be objectively reasonable based on the facts and circumstances perceived and/or known by officers at the time the force is used. Officers should take into consideration: the severity of the crime involved; the actions of the subject and/or third parties; whether the subject poses an immediate threat to the safety of officers or others; and officer/subject factors. Officers' decisions are often made in circumstances that are stressful and rapidly evolving, thus officers are not required to determine the least intrusive measure of force that might resolve the situation. Officers are required to select an objectively reasonable option based on the perception of the risk and the actions of the subject and/or third parties at the time, as well as drawing upon their own training and experience.

2. When deciding on what specific force response to use, a balance between the force response chosen and the potential/foreseeable risk of injury to a subject are factors that should be considered in assessing the

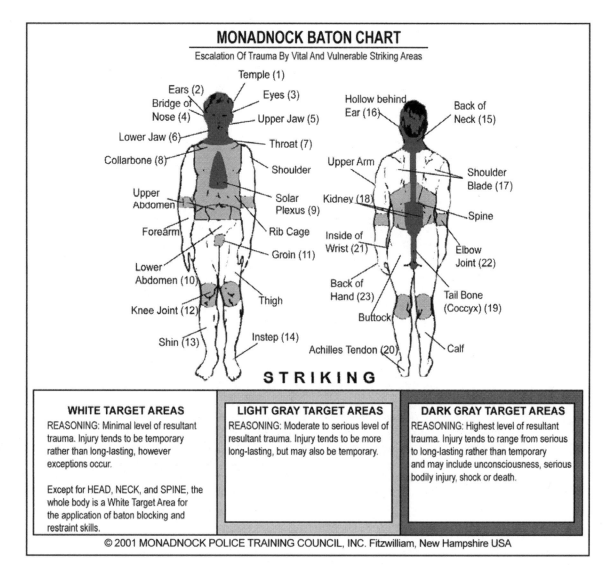

Figure K-1. Escalation-of-Trauma Chart

reasonableness of that use of force. When officers decide that a baton technique is an objectively reasonable option, they would then need to choose an appropriate target area. Target selection based on medical evaluations of the vulnerability of various parts of the human body and potential injury to subjects may assist officers in reducing injuries to subjects. In the Monadnock Baton Chart, the potential trauma to the human body has been designated by shades of gray denoting the level of risk incurred by the application of physical force by means of the baton.

K.4.4 Background

The concept of white, light gray, and dark gray target areas was developed to assist officers in assessing the probability of injury to subjects. When time allows, officers' use of force should take into consideration escalating and deescalating options based on threat assessment, officer/subject factors and the probable severity of injury.

K.4.5 The Concept in Action

White target areas are considered primary target areas when a baton technique such as a strike is chosen. Light gray target areas are considered an option when force applied to a white target fails to overcome resistance or does not correspond with the threat level. Gray targets are areas of the body where force is directed at a joint or an area in close proximity to a prominent dark gray target area and therefore the risk of injury increases. Dark gray target areas are for confrontations where the subject is attempting to cause serious bodily injury or is applying deadly force to an officer or another; or situations where force to lower level target areas is ineffective based on an escalation of resistance presented by the subject during an attempt to end the confrontation. Physical force directed at dark gray target areas poses a greater risk of injury to the subject and in certain areas might constitute deadly force because of the probability of causing death.

The general policy of the Monadnock Police Training Council is stated in the Note below.

Note

"The Monadnock Police Training Council would support an agency or instructor who finds it necessary to raise a specific target area to a higher color-code classification; for example, the collarbone from a light gray to a dark gray target area. Any agency or instructor who elects to do so must clearly document that action. The Council would not support any change whereby a specific target area's color-code is lowered; for example, changing the collarbone from light gray to a white target area. As a reminder, force applied responsibly always takes into consideration the consequences of actions and focuses on the ultimate goal or purpose of action."

K.5 RIOT CONTROL AGENTS AND OLEORESIN CAPSICUM SPRAY

1. NSF may use RCAs to subdue a subject when used in self-defense or protection of a third party if circumstances warrant their use.

2. The Navy shall purchase and use only OC sprays that are nonflammable and noncarcinogenic.

3. The NSF shall purchase and use only OC sprays within the following ranges of potency: 5–10 percent OC, 0.8–0.22 percent capsaicinoid content level, and a Scoville heat unit (SHU) rating of up to 50,000. There are three methods of identifying the potency of OC spray. The methods in (a) and (b) are imprecise unless the effective ingredient of the product that is measured is pure capsaicin. The method in (c) should be determined by high-pressure liquid chromatography:

   a. OC content by percentage of oleoresin capsicum.

   b. Capsaicinoid content level.

   c. Heat rating in SHUs.

4. Navy activities should purchase OC sprays that:

   a. Use propellants that maintain constant pressure until the canister is drained.

   b. Provide material-safety data sheets listing materials contained in the spray and known hazards.

   c. Use a propellant system that is safe for the environment and effective in the manner in which it is to be used (fog, stream, mist, or foam).

   d. Do not contain other chemical agents such as CS or CN.

5. Activities should ensure that OC spray canisters are replaced after five years to ensure that the canister retains full pressure.

6. If RCAs are used, personnel shall receive medical attention as soon as possible. In the absence of prompt medical assistance, NSF personnel involved in the administering of OC spray shall be responsible for

seeking medical attention for the offender(s) exposed to OC spray as soon as practical. It is not mandatory and usually not necessary to afford subjects immediate medical attention. People who have been sprayed shall not be released until they have been advised of the safety measures to be taken or until medical treatment has been received.

7. NSF personnel shall not be issued chemical agents until they have been trained in the use of the agents and have demonstrated a knowledge of safety precautions involved with chemical agents, including necessary medical treatment following its use. Training will be IAW the INIWIC. Initial entry civil service and contract employees (recruits) shall be processed to Level 1 OC contamination during the initial training phase. This requirement shall be established in their position description, and the employee shall accept this training as a condition of employment. Refusal to undergo Level 1 OC training is grounds for dismissal. Additionally, all NSF, including contract personnel, who are issued OC spray during the performance of their duties shall undergo approved Level 2 or 3 training annually. Training documentation must specifically describe the level of training received. The rationale is that direct contamination is necessary for several reasons, to wit:

a. Likelihood that an officer will inadvertently contaminate himself or that another officer will directly or indirectly contaminate him during a high-intensity struggle to control a combative subject. If the effects of OC have not been personally experienced by the officer using the product, the intended symptoms that overcome a combative subject will overcome the officer in most cases. The result is panic and potentially use of deadly force, with a self-defense justification offered as a defense to this action.

b. Weapons retention during a struggle where direct or indirect contamination has occurred is specifically addressed through performance exercises. Historical experience indicates that first-time contamination with OC during these drills often results in the officer attempting to withdraw from the struggle or altogether quitting due to the unfamiliar sensation of burning in the soft mucous membrane areas of the face. However, students who gain experience with this effect often control their response much better, use less force during the event, and proceed with the understanding that the effects are temporary.

8. Manufacturer's instructions combined with the direction outlined herein should be followed whenever employing OC spray.

9. Use OC spray only with consideration of all relevant circumstances, such as wind direction and speed; bystanders' presence; and whether the suspect is operating a motor vehicle, on a rooftop, or in another situation where the subject's debilitation from OC spray would cause an unreasonable danger to self or others.

10. Do not use OC spray in an enclosed environment, such as aboard an aircraft or automobile. Extreme caution should be used when considering use within an interior compartment aboard a ship. DOT regulations prohibit the carrying of OC spray aboard aircraft.

11. Use OC spray against aggressive or threatening animals only when no other reasonable alternative is available, providing the amount used is as humane as possible and fitting the circumstances.

12. Use of OC spray on foreign installations is subject to SOFAs with HNs. Commanders should consult their SJA for clarification on any issues pertaining to location (in or out of the continental United States) and ensure the use of OC spray is not in violation of existing local laws or agreement.

13. During peacetime, RCAs may be used on U.S. bases, posts, embassy grounds, ships, and installations for protection and security purposes, riot control, and evacuation of U.S. noncombatants and foreign nationals. The U.S.-controlled portions of foreign installations are considered U.S. installations.

14. NSF LE personnel may use OC spray for the performance of LE activities:

a. On base (CONUS and OCONUS).

    b. Off base when authorized by exception to Posse Comitatus Act (18 U.S.C. 1385) in the United States and its territories and possessions.

    c. Off base overseas in those countries where such use is specifically authorized by the HN government.

    d. Off base (worldwide) for the protection or recovery of nuclear weapons under the same conditions as those authorized for the use of lethal force.

    e. In training.

15. The employment of OC spray is coupled with the requirement to provide first aid to the subject sprayed with OC when reasonable.

16. Commands will report any OC-attributed adverse effect sufficient to require treatment by a medical provider to Chief of Naval Operations (N314).

K.6 MILITARY WORKING DOGS

Per OPNAVINST 5585.2 (series), Department of the Navy Military Working Dog (MWD) Program, MWDs trained to attack (i.e., patrol, patrol/drug detector, and patrol/explosive detector dogs) must be considered a weapon.

1. When releasing an MWD to attack, the handler must:

    a. Be sure that the MWD will cease an attack upon command.

    b. Be sure that the MWD has identified the same target that the handler is releasing it to attack.

    c. Warn bystanders to cease all movement.

    d. Call the MWD off the attack as soon as the suspect stops resisting or indicates surrender.

2. MWDs will not be released to attack:

    a. If no suspect is in sight.

    b. In areas where children are present except as a last resort short of the use of a firearm.

    c. Into a crowd of people.

K.7 SUMMARY

NLW capabilities can provide a forgiving means of imposing our will on adversaries when deadly force is not the preferred force option. The proper applications of force capabilities, lethal and nonlethal, are critical components to any successful military operation. In the contemporary operating environment, a key aspect for mission accomplishment is public support. To defeat our adversaries, U.S. forces must be able to separate enemies from innocent people. At the same time, U.S. forces must conduct themselves in a manner that enables them to maintain popular domestic support. The lack of NL capabilities and excessive or indiscriminant use of force are likely to alienate the local populace, thereby increasing support for the enemy. Insufficient tools and applications of force result in increased risks to friendly forces and perceived weaknesses that can jeopardize the mission by emboldening the enemy and undermining domestic popular support. Achieving the appropriate balance requires a commander to have a thorough understanding of the tools available to his/her unit, the policy under which those force options are authorized, and the nature of the enemy.

APPENDIX L

Arrest/Apprehension/Detainment

L.1 DEFINITIONS

1. Apprehension is the taking of a person into custody. Apprehension is the equivalent of arrest in civilian terminology. An apprehension is different from detention of a person for investigative purposes, although each involves the exercise of government control over the freedom of movement of a person. An apprehension must be based on probable cause, and the custody initiated in an apprehension may continue until proper authority is notified and acts under MCM Rule 304 or 305.

2. An investigative detainment may be based on less than probable cause and normally involves a relatively short period of custody. Furthermore, an extensive search of the person is not authorized incident to an investigative detention, as it is with an apprehension.

3. An apprehension is the restraint of a person by oral or written order not imposed as punishment, directing the person to remain within specified limits.

L.2 APPREHENSION AUTHORITY

1. NSF members are representatives of the U.S. Government, the U.S. Armed Forces, and, when assigned to an installation, the CO. MCM Rule 302(b)(1) gives NSF the authority to apprehend individuals.

2. NSF may apprehend any person subject to the UCMJ if they have a reasonable belief the person being apprehended has engaged or is engaging in criminal activity. A reasonable belief is a belief based on the kind of reliable information that a reasonable, prudent person would rely on, which makes it more likely than not that something is true. For personnel subject to the UCMJ who have been apprehended and/or detained, disposition procedures shall be IAW local policy established by the CO.

3. NSF have limited authority to apprehend persons not subject to the UCMJ. In areas under military jurisdiction or control, NSF may take persons not subject to the UCMJ into custody:

   a. Who are found committing a felony or misdemeanor amounting to a breach of peace on a military base. Such persons must be turned over to civil authority as soon as possible.

   b. Who are violating properly promulgated post regulations. These persons may be escorted to the entrance of the base and may be forbidden reentry by the base commander as necessary. If counterintelligence or terrorist-related activities are suspected, NCIS must be immediately notified prior to release.

   c. In some cases, persons not subject to the UCMJ may be cited for violations of the Assimilative Crimes Act (18 U.S.C. 13) not amounting to felonies or breaches of the peace and referred to a U.S. magistrate.

L.3 DETENTION OF CIVILIANS

Title 18, U.S.C., and the U.S. Constitution authorize the detention of civilians for offenses committed on a military installation. Since civilians are not normally subject to the UCMJ, refer civilian violators to a U.S. magistrate for judicial disposition or to the local civil authorities having jurisdiction. For minor offenses, release civilian offenders to their military sponsor. If they do not have a military sponsor, release them to a relative or friend or on their own recognizance.

L.4 OFF-INSTALLATION PATROLS

NSF performing patrol duties off the installation have the authority to apprehend military personnel if authorized by the CO. However, a civilian police officer should be present to identify all suspected military violators in civilian clothing.

L.5 OVERSEAS AND HOST-NATION POLICY

The CO may authorize off-installation patrols if legal and HN coordination has been accomplished. NSF have the authority to apprehend military personnel on or off the installation in an overseas environment. The authority to detain civilians on a U.S. military installation varies in each HN. Bilateral agreements and directives must specify such limitations. The SJA will provide guidance.

L.6 CUSTODY

Custody is the restraint of free movement. An apprehension occurs when a NSF member tells a suspect he/she is under apprehension. Once suspects are apprehended and in the custody of the person who affected the apprehension, NSF not only controls their action and movement but are also responsible for their safety.

L.7 SEARCHES

Immediately upon apprehending a suspect, conduct a search of the suspect for weapons and any evidence that the suspect could remove or destroy. This step emphasizes the safety of the security police and the apprehended individual. (See Appendix H of this NTTP for further guidance on searches.) The apprehending NSF makes the decision to search without handcuffs or to search handcuffed and, based on the situation at hand, may also search the area under the suspect's control.

L.8 USE OF FORCE

During apprehension and detention of suspects, always use the minimum force necessary. Use handcuffs, chemical irritant projectors, batons, and firearms only when specifically trained in their use. In addition, NSF must strictly adhere to approved employment procedures. (See Appendixes B, C, H, and T of this NTTP for specific TTP.)

L.8.1 Handcuffs

1. The courts consider handcuffing a use of force; therefore, NSF must carefully analyze each situation to ensure they use the minimum level of force to protect themselves and others from injury. During an apprehension, for the safety of the NSF, apply handcuffs to ensure control of the apprehended individual during detention and search, at the apprehension site, and during transport. (See Appendix T of this NTTP for specific TTP.)

2. When applying handcuffs, security police use a reasonable level of force to achieve control of a resisting detainee. Inform unresisting detainees of the handcuffing procedure and give them the opportunity to cooperate.

3. The use of handcuffs is a precaution against an apprehended person who may become uncooperative or violent, to prevent escape, or to ensure personal safety.

L.8.2 Chemical and Inflammatory Agents

Use authorized chemical agents as an intermediate-level use of force within the United States and its territories and possessions.

L.8.3 Expandable Baton/Riot Baton

Batons may be used if NSF are trained/qualified and the batons are authorized at the installation. If it is necessary for normal wear, then consider the unobtrusive, collapsible type.

L.8.4 Firearms

The NSF routinely bear firearms in the performance of duties. When responding to an incident where they may meet an armed adversary, they must be prepared to use deadly force. (See Appendix B for TTP.)

L.9 MIRANDA RIGHTS AND THE UNIFORM CODE OF MILITARY JUSTICE

1. Both UCMJ Article 31 (for military personnel) and the Fifth Amendment to the U.S. Constitution (for civilian personnel) protect an individual from being compelled in any criminal case to be a witness against him/herself.

2. Specifically, a person subject to the UCMJ who is required to give warnings under Article 31 may not interrogate or request any statement from suspects or persons suspected of an offense without first advising individuals of their rights.

3. NSF may question an individual without a warning up to the point where they suspect the individual may have committed a crime.

   a. NSF may ask an individual the following administrative questions without providing a warning: name, social security number, duty station and/or address, date of birth, height, rank (if military), weight, etc.

   b. An appropriate warning must precede any incriminating questions.

4. The following special provisions apply for rights advisement:

   a. The warning should be given by LE or investigative personnel who intend to question an individual suspected of an offense.

   b. The point at which an individual is suspected of an offense will be determined by the facts of the situation. In general, whenever there are sufficient facts to indicate that a crime has occurred and that a particular individual may be culpable, then a warning shall be given.

   c. The determination of when an individual is in custody has been the subject of several Supreme Court decisions, and a specific determination of this matter should be obtained from the local SJA.

   d. Military Suspect's Acknowledgement and Waiver of Rights (OPNAV 5580/3) will be used to document the military subject's waiver of rights. This form also provides space at the bottom to begin a written statement. A written statement must always begin on the same page as the waiver of rights.

   e. Civilian personnel.

      (1) Civilian personnel suspected of an offense should be provided with a Miranda warning if they are in custody or otherwise deprived of freedom of action in any significant way.

      (2) The point at which a civilian is suspected of an offense is the same as for military personnel.

      (3) The warning must be given by LE or investigations personnel.

      (4) Civilian Suspect's Acknowledgement and Waiver of Rights (OPNAV 5580/4) will be used to document the civilian subject's waiver of rights. This form also provides space at the bottom to begin

a written statement. A written statement must always begin on the same page as the waiver of rights. No locally produced forms may be used in place of OPNAV 5580/4.

(5) A civilian warning must be given to a juvenile in terms that he or she can understand. The warning should be given in the presence of a parent or guardian, if at all possible. If a parent or guardian is not present, ensure that an advocate is present while the juvenile is advised of his or her rights.

f. Foreign countries.

(1) In most foreign countries where the United States maintains military facilities, foreign nationals (citizens of that country or another foreign country) who commit an offense against the property of the United States or against the person or property of members of the Armed Forces located at the activity are not subject to the laws of the United States. However, with respect to U.S. Naval Station Guantanamo Bay, which is part of the Special Maritime and Territorial Jurisdiction of the United States as defined at 18 U.S.C. 7(3), foreign nationals can be prosecuted for violations of U.S. law committed there.

(2) Suspects who are foreign nationals should be warned or advised IAW the procedures that control such advice in the country where the base is located. Such situations are extremely sensitive, and specific guidance shall be obtained from the local SJA before any process of apprehension occurs. SOFAs may also apply.

(3) Do not question non-English-speaking U.S. citizens or foreign visitors until their level of understanding of their rights can be fully ascertained. Call the SJA for guidance in any cases where there are questions.

5. When a suspect has made incriminating statements in response to questioning by NSF, a command representative, or other personnel during questioning/interrogation without a warning of rights beforehand, or when it is suspected that an existing criminal admission was improperly obtained from a suspect, the suspect must be advised that the previous admission cannot be used against him or her in a trial by court-martial or other court of law. A "cleansing warning" must be provided to the suspect (e.g., "I advise you that any prior admissions or other improperly obtained evidence which incriminated you cannot be used against you in a trial by court-martial or other court of law"), as appropriate. But in the case of an unsolicited statement or spontaneous utterance (i.e., not in response to interrogation or questioning), a cleansing warning would not be required or even desirable (from the government's perspective). This statement will be acknowledged at the bottom of the appropriate rights advisement form. Prior questionable admissions made by a suspect must be completely documented, to include particulars about previous warnings. This information shall be included in the investigative report and any documents pertaining to the report.

6. NSF may ask questions of a suspect without a warning for the NSF's personal security if the suspect may have information relative to the scene of the incident, knowledge of which could protect the NSF member.

7. A witness or victim of a crime will not be provided with a military or civilian warning of rights. If, during an interview, a witness or victim is reasonably suspected of an offense or of making false statements, then appropriate warnings should be given at that time.

L.10 TRANSPORTING APPREHENDED OR DETAINED PERSONS

All persons in custody shall be searched for weapons before being placed in a vehicle and transported; place them in the vehicle to allow the use of installed seatbelts. When transporting a suspect, NSF must provide dispatch with the departure time and destination arrival time (along with odometer readings).

APPENDIX M

Domestic Violence/Crisis Intervention

M.1 DOMESTIC VIOLENCE

Family violence is a serious national problem. NSF members can expect to encounter potentially violent incidents of domestic violence. NSF personnel need to know how to defuse these incidents and regain control of the situation. Additional information can be found in DODD 6400.1, Family Advocacy Program (FAP).

Responding to family violence is a difficult, demanding, and sometimes dangerous job. Situations of family violence are inherently dangerous. The risk of injury to a family member or the responding NSF is always present. Each year LE personnel are injured or killed responding to family violence calls.

Studies have shown that family violence calls are often repeat calls. When the initial call is not effectively handled, it is likely the situation will recur and the patrol will be called again. By learning how to recognize child abuse/neglect and spouse abuse and by acquiring the skills necessary to act accordingly, NSF personnel can avoid repeat calls, provide protection to victims of family violence, and reduce the likelihood of injury to the family member.

NSF personnel have a critical role to play both in restoring order and in preventing future incidents of family violence. NSF are available 24 hours a day, seven days a week. They are trained to respond in an emergency and have transportation and communication equipment immediately available. NSF also have the authority to intervene and, if necessary, to apprehend to restore order and protect lives.

M.2 DEFINITIONS

In discussing family violence, the following definitions will be used:

1. Child: An unmarried person, whether a natural child, adopted child, foster child, stepchild, or ward, who is a family member of the military member or spouse and who either:

   a. Has not passed his/her eighteenth birthday; or

   b. Is incapable of self-support because of mental or physical incapacity that currently exists and for whom treatment is authorized in a medical facility of the uniformed Services.

2. Spouse: A partner in a lawful marriage.

3. Abuse: Physical injury or emotional distress inflicted by other than accident, as evidenced by but not limited to scratches, lacerations, skin bruising, bleeding, malnutrition, sexual molestation or abuse, burns, bone fractures, subdural hematoma, soft-tissue swelling, and unexplained death or where the history given concerning such a condition is at variance with the degree or type of the condition or where circumstances indicate the condition may not be the product of an accidental occurrence.

4. Child neglect: Neglect is failure to provide a child with the basic necessities of life. The lack of medical care, inadequate nourishment, insufficient clothing, lack of supervision, and unsafe living conditions may indicate neglect. However, care must be taken in differentiating willful neglect from poverty. Impoverished families may be providing the best care possible within their means.

5. Navy Family Advocacy Program: The Navy-wide program developed and designed to identify, prevent, and treat consequences of child abuse and neglect and spouse abuse and to provide other necessary medical and nonmedical services for victims and perpetrators of child abuse or neglect and spouse abuse.

6. Child Advocacy Program: That part of the Navy Family Advocacy Program that deals with child abuse or neglect.

M.3 SPOUSE ABUSE

Spouse abuse includes wife battering, husband battering, and simultaneous fighting in which either or both spouses may be assaulted.

M.3.1 Types

Three types of spouse abuse are generally recognized. All types are dangerous. Even a single episode must be taken seriously because it may result in severe injury or death.

1. Physical battering, which includes punching, kicking, beating, stabbing, shooting, and other forms of physical assault.

2. Sexual battering, which includes physical attacks on the victim's breasts or genitals and forced sexual activity.

3. Psychological battering includes harassment, threats (e.g., suicide threats or threats to harm the children), emotional maltreatment, and deprivation of food and sleep. Another example of psychological abuse could include the destruction of property, pets, or both of value to one spouse—a beloved dog, a car, or family heirloom.

M.3.2 Characteristics

Spouse abuse occurs among all ethnic, racial, age, and economic groups. It is frequently linked to alcohol abuse. Spouse abuse tends to be recurrent, increasing in severity and intensity unless there is appropriate, timely intervention by those trained to work with both the aggressor and the victim.

M.4 CHILD ABUSE AND NEGLECT

Like spouse abuse, child abuse and neglect may be found in families of all ages, race, and ethnic backgrounds and at all economic levels. Victims may be of any age from infancy to adolescence, but the majority are of school age. However, children under the age of three are most likely to be severely injured by the abuse.

M.4.1 Types

Three types of child abuse and neglect are recognized. They often occur in combination. Thus, a physically abused child also may be emotionally maltreated; a sexually abused child also may be neglected. Sometimes one child in a family is singled out for abuse or neglect; this child becomes the family scapegoat. In other families, all of the children may be abused or neglected. Children of any age may be abused or neglected.

1. Physical abuse includes nonaccidental injuries such as burns, spiral fractures of the arms or legs, and bruises in various stages of healing. Sometimes a pattern of injuries can be discerned (e.g., bruises on both sides of the face or a glove- or sock-like burn). Sometimes the mark of the implement used to injure the child can be seen (e.g., the imprint of a belt buckle, looped electrical cord, or a cigarette burn).

2. Sexual abuse includes, but is not limited to, sexual intercourse; sexual contact, such as fondling of the genitals; exhibitionism or masturbation in the presence of the child; and sexual exploitation, such as child pornography. The sexually abused child may be a boy or girl of any age, but it is frequently a girl of four to nine years of age at the onset of the sexual abuse, which then continues into adolescence. Sexual abuse

most commonly occurs between father and daughter, but father-son, mother-daughter, mother-son, sister-brother and family friend-child abuse also occurs.

3. Emotional maltreatment includes ignoring, threatening, terrorizing, or blatantly rejecting the child. Emotional maltreatment is the most difficult type to identify. There are few physical indicators to serve as guideposts. Emotional maltreatment is normally identified by trained professionals who examine the behavior of both parents and child.

M.4.2 Indicators

The indicators of child abuse and neglect may be physical or behavioral. Indicators may be seen in the child, parent, or interactions between the parent and child. Abuse may be revealed through the child's school, child care center, routine physical examinations, or hospital visits. No one indicator by itself proves child abuse or neglect is occurring. However, the presence of several indicators together should alert the NSF to the possibility of child abuse and neglect.

1. Physical indicators include burns, bruises on several different surfaces of the body (in the genital area, over large areas of the body or in various stages of healing), human bite marks, hunger, venereal disease and pregnancy, especially in a child under the age of 13.

2. Behaviorally, maltreated children may be wary of adult contact. They may appear fearful, withdrawn, frightened, or hungry. The sexually abused child may have difficulty walking or sitting. The neglected child may beg or steal food.

3. Abusive parents are frequently immature, dependent people who are rigid disciplinarians and have unrealistic expectations of their children. They show little concern for their children or may offer conflicting, improbable or no explanation for their children's condition. They may fail to look at, touch, or comfort their children and may expect the children to look after them. In many cases these parents were abused or neglected when they were children.

4. Abusive families are sometimes families under stress. The stress may be financial, personal, job-related, environmental, or situational. In any event, it is a situation with which the parent cannot cope. Like spouse abuse, child abuse and neglect may be recurrent, increasing in frequency and severity unless the family receives help.

M.5 NAVY SECURITY FORCES' ROLE IN DOMESTIC VIOLENCE

One of the missions of NSF is to maintain law and order within the command. To reestablish order and preserve the peace, NSF personnel are often required to respond to situations of family violence. The primary role in these situations is to take immediate action to restore order and protect lives. Responding NSF should avoid taking sides with either party in the dispute. In addition, the sponsor's commanding officer, CDO, or command representative and the base family advocacy officer must be advised of all incidents of family violence. NCIS is responsible for investigating major offenses, including child abuse and spouse abuse involving aggravated assault. The local NCIS office must be notified immediately of all child abuse or neglect allegations. NSF routinely investigates minor incidents of child abuse/neglect and spouse abuse. Incidents shall be recorded on the DON Incident Report/Addendum-Victim (OPNAV 5580/1D) and DON Desk Journal (OPNAV 5580/19) within CLEOC. When peace has been restored and the appropriate authorities notified of the incident, the NSF member's role has ended except where an offense has been committed and an apprehension will take place.

M.6 COMMANDING OFFICER'S ROLE IN DOMESTIC VIOLENCE

Commanding officers are responsible for the actions of assigned personnel, both on and off duty. This responsibility includes the behavior of military sponsors and their military family members within military housing areas and in economy housing overseas where authorized. NSF members should recognize this command responsibility and duty to inform the subject's commanding officer through the CDO or designated command representative of incidents of family violence involving unit personnel and their family members. The commanding officer or designated representative is responsible for initiating military protective orders.

M.7 FAMILY ADVOCACY OFFICER'S ROLE IN DOMESTIC VIOLENCE

The family advocacy officer functions as the central coordinator for the base FAP. Under the direction of the director of installation medical services or chief of hospital services, the family advocacy officer performs a number of family violence prevention and intervention services.

M.8 DISPATCHER'S ROLE IN DOMESTIC VIOLENCE

In answering family violence calls, dispatchers must have all available information on the family and situation. The dispatcher has the responsibility for obtaining as much information as possible from the individual making the call while dispatching a patrol unit. When a possible family violence call comes in, the dispatcher:

1. Should obtain as much data as possible from the caller, asking who is involved, what has happened, when it occurred, where it happened and where the family is now, how the incident occurred, whether weapons are involved or available, and whether medical aid is required.

2. Ask the caller to describe the situation and for the telephone number from which the call is being made in case the call is disconnected. The dispatcher should try to keep the caller on the telephone line until the arrival of the responding NSF.

3. Communicate all information to NSF responding to the incident. If the dispatcher is unable to obtain a clear description of the situation, the responding unit(s) must be so informed.

4. Check locally devised firearms registration roster/database and whether a previous military protective order has been issued.

5. Ensure an appropriate patrol response with follow-up actions and referrals to commanding officers through the CDO or designated command representative, family advocacy officer, or NCIS. In cases where the reported abuse is not currently ongoing, a patrol response may not be appropriate. In these instances, coordination between the dispatcher and the supervisory chain of command, the commanding officer of the military member involved, the NCIS, and the family advocacy officer will determine the appropriate response.

6. In cases of notification by the installation hospital of a suspected child abuse incident being treated, NCIS should be contacted immediately to meet the dispatched patrol.

M.9 PATROL RESPONSE

M.9.1 Arriving Safely

The responding patrol(s) should drive to the scene as quickly and safely as possible. The responding patrol(s) should formulate a general plan before arrival. It is wise to determine in advance who will be in charge, who will approach first, who will serve as backup, etc.

Use discretion when approaching a house where a domestic disturbance is occurring. Park NSF vehicles at least one house away from the address of the incident. Parking directly in front of the house in question may escalate the situation or warn residents to conceal evidence of a crime. For the same reasons, flashing lights and sirens should be off before arrival, and there should be no loud noises like slamming car doors or radios.

M.9.2 Approaching the Scene

NSF members should use caution when approaching the scene of a family violence complaint. Before approaching the house, NSF members should stop, look, and listen. Windows, doors, adjoining buildings, or areas of possible concealment should be visually checked for unusual movements or objects. The NSF members should not consider approaching the side or back door instead of the front door, as they could be mistaken for prowlers. If the approach is made at night and flashlights are used, they should not be shined in windows. Avoid silhouetting the other patrols.

M.9.3 Entry Procedures

1. NSF members should always be dispatched in pairs to family violence calls and should always stand to one side of the door, never in front of it, and avoid standing in front of windows if possible. The second NSF member should be behind and to one side of the first, in position to maintain visual contact with the inside of the residence and provide cover.

2. Before knocking, listen at the door for 15–30 seconds. NSF members may be able to obtain information on the nature of the disturbance and whether or not it is violent before announcing their presence.

3. Check screen doors before knocking to see whether they are locked. Locked screen doors can create an unexpected barrier between the NSF member and residents if immediate action is required.

4. Knock on the door in a reasonable manner, if at all possible, or use the doorbell.

5. Evaluate the risk of entry, even when invited to enter, and respond accordingly.

6. If there is no response at the door and the dwelling appears quiet, the address should be verified with the dispatcher. If the address is correct, the sides and rear of the quarters should be checked for indications of the presence of the occupants. Neighbors may also provide useful information.

M.9.4 Initial Contact with Residents

1. NSF members should display a calm, positive, and helpful manner. Initial impressions will set the tone for the interview.

2. When someone answers the door, NSF members should introduce and identify themselves and state the reason for their presence.

3. If not invited into the dwelling, NSF members should request to move the discussion inside. This will remove the situation from the view of the neighbors and enable NSF members to observe:

    a. Any injuries requiring treatment

    b. The location and number of the disputants

    c. Visible weapons and threatening moves

    d. Living conditions

    e. Emotional stage of dispute and emotional condition of disputants

    f. Impairment

    g. Children at risk

    h. Physical damage to property.

4. Separate the disputants as necessary and maintain visual contact with the other NSF member.

5. After providing for any necessary medical assistance and calming the situation, obtain information on the family structure and background. Such information will give important background and data and allow a cooling-off period. Questions asked may include:

   a. Names and whereabouts of the sponsor and family members

   b. The sponsor's (and any other military member's) rank, social security number, and unit

   c. Relationship and legal status of residents (e.g., nephew, uncle, boyfriend, girlfriend, valid marriage)

   d. Length of residence in quarters and period assigned to installation

   e. Ages and relationships of children

   f. Whether military or civilian police have been required to respond to previous incidents

   g. Whether the family has been to a family advocacy office.

6. Alcohol is involved to some extent in many situations of family violence. If one family member is intoxicated, it may be difficult to obtain factual information from that person. Other sources of information may have to be obtained.

M.9.5 Visual Observation

Observing conditions inside the quarters while obtaining background information may give ideas of the cause contributing to the situation. The behavior of residents can provide important clues.

1. Signs of fear, hate, depression, and embarrassment can be detected in facial expressions, eye movements, and body positions.

2. Be alert for sudden movements and continual glances at closed doors, closets, or bureaus. Such actions may be the first indication the individual has a weapon available or is attempting to conceal the presence of an injured family member or other evidence.

3. The condition of the home and appearance of the residents may provide clues to family functioning. If the living conditions are unusual, unsafe, or unhealthy, NSF members may want to arrange for photographs of the scene and request response of the military member's commanding officer or CDO.

M.10 TYPES OF DISPUTES

M.10.1 Violent Disputes

1. When responding to a violent disturbance, NSF members must immediately separate the disputants. If medical attention is required, it should be secured at once.

2. NSF members should be vigilant about their personal safety as well as that of disputants. In separating the persons involved, make a visual search for objects that could be used as weapons. The disputants should never be allowed to come between the NSF members. Disputants should never be left alone in another room and should not be removed to the kitchen because of the availability of potential weapons. If the disputants cannot be calmed, apprehension and removal to the security department may be necessary.

3. Intoxicated people tend to be violent in disputes. It may be extremely difficult to reason or deal with a person in this condition. The individual may have to be removed from the scene until sober enough to be reasoned with.

4. A potential danger exists in persons who are unusually quiet and controlled in highly emotional disputes. Such people may be near the breaking point and may become violent and upset by an innocent gesture or remark.

5. If the parties can be separated and NSF members are not at increased risk, the disputants may be removed out of sight and hearing of each other. Once they are separated and order is restored, the parties may be interviewed.

M.10.2 Verbal Disputes

The difference between violent disputes and verbal disputes is that in a verbal dispute, a physical assault has not occurred. The parties involved may be easier to reason with and prompt solution to the dispute more possible.

1. Remove the disputants to separate rooms if possible. Avoid leaving them alone or in the kitchen.

2. Separating normally causes a distraction to the disputants. NSF members' use of a calm, firm, and assured tone of voice may further distract the disputants and better control the situation. Once they are separated and order is restored, the parties may be interviewed.

3. In disputes where one disputant is a child or young adult, the youth may resent authority figures and assume that NSF will automatically side with the parents. Therefore, when answering such a disturbance call, an attitude of concern and understanding for the child's version of the argument is important. The youth's feelings, problems, and thoughts should be listened to and evaluated as carefully as those of the parents or other disputants. However, take care not to interfere with parental rights regarding the children. If other children are present but not involved in the dispute, the parents should be asked to remove them from the room.

M.11 ATTITUDE OF RESPONDING NAVAL SECURITY FORCES

The attitude of the responding NSF is important and, in many cases, will determine the attitude and cooperation of the family members involved.

1. In responding to a call, keep in mind that each situation is different and must be treated individually. Meanings and attitudes might be read into words, facial expression, and body positions. The best approach is a calm and positive one.

2. Realize that people may be hostile, angry, frightened, ashamed, or uncooperative when LE personnel appear at their door. They may view the NSF as intruders and resent their presence. Carefully consider opening remarks and questions. The NSF member's approach should be calm, controlled, and concerned. Care should be taken to avoid sarcastic or critical remarks, an impolite tone of voice, or threatening or aggressive body positions. Hostility, indifference, or aggression may provoke further violence, while a sensitive and tactful approach may restore order and calm the situation.

M.12 THE INTERVIEW

Once the parties have been separated, each person should be interviewed to:

1. Assess the immediate danger to family members and need for medical assistance or protective custody.

2. Determine whether suspected abuse or neglect is occurring.

3. Determine the appropriate response to the situation.

4. Identify the perpetrator or primary aggressor.

5. Protect the legal rights of suspects.

6. Many times the aggressor will have marks, such as scratches, which are often caused by the victim through defensive measures.

7. Identify victims and give them proper assistance.

8. Interview any/all children who are present at the scene whenever possible.

M.13 NAVY SECURITY FORCE ACTIONS

After order has been restored in family violence situations, three actions may occur: referral, temporary separation, or apprehension. One or more of these actions is possible.

1. Referral. If the dispatcher has not already done so, request the unit commanding officer or CDO, NCIS, and family advocacy representative be notified of the situation, as appropriate. The family advocacy representative must be notified of all incidents or complaints involving child or spouse abuse. The NSF member's role in this instance is to maintain order, secure the scene as necessary, and stand by until relieved or advised of further responsibility for the case.

2. Temporary Separation. Family members may be separated to ensure safety and protection. For example, the CDO may suggest that a spouse temporarily leave the quarters, or a commander may order the separation by restricting the military member to a dormitory or military lodging facility. The hospital may admit a child for medical treatment or observation when medically appropriate. If a child's safety is threatened, take appropriate action under state law, including contacting child protective services. NCIS has primary jurisdiction for serious child abuse or neglect involving infliction of serious bodily harm. The NSF member's role in this instance is to provide whatever support is requested by the hospital or commander.

3. Apprehension. Apprehension may be the required course of action when:

   a. There is a formal complaint.

   b. Probable cause exists that a violation of the UCMJ or local law has occurred.

   c. The family member refuses to cooperate with NSF, CO, CDO, or family advocacy officials.

In all instances involving an apprehension, an OPNAV 5580/1 within CLEOC must be completed detailing the situation, NSF response, and the parties responsible for follow-up actions.

NSF personnel investigating domestic abuse/violence must ensure victims and witnesses are provided a DD Form 2701, Initial Information for Victims and Witnesses of Crime, to ensure victims/witnesses are aware of their rights under the Victim and Witness Assistance Program. Ensure the DD Form 2701 has been annotated in the DON Incident Report/Addendum–Victim (OPNAV 5580/1D) and the DON Desk Journal (OPNAV 5580/19).

APPENDIX N

Law Enforcement Response in an All-Hazards Environment

N.1 OVERVIEW

NSF LE responses occur in an all-hazards environment. The NSF must be prepared to respond to the range of all foreseeable hazards, including natural and man-made disasters, such as hurricanes, earthquakes, forest fires, floods, leaks at chemical plants, oil spills, radiological contamination, and power outages; nuclear, biological, or chemical attack; and sabotage, including attacks against critical infrastructure. For more in-depth guidance regarding response in an all-hazards environment, refer to CNICINST 3440.17, Navy Installation Emergency Management (EM) Program Manual.

The actions of the first NSF on the scene set the stage for the remainder of the response, including the safety of all responding personnel. The first priority must be self-protection. NSF responding to a WMD incident or an unexplained explosion must make an assessment on arrival at the incident scene. Secondary devices must be taken into account. NSF should take several immediate steps to protect themselves after observing abnormal circumstances, such as unexplained dead or dying animals, multiple casualties, smoke cloud, unexplained fluids, etc. NSF should be able to perform initial on-scene management effectively until relieved by higher authority. The following list outlines actions that are necessary to gain initial control of the response effort.

N.2 APPROACH TO THE SCENE

1. Immediate actions.

    a. Protect self.

        (1) Approach from upwind.

        (2) Maintain safe distance (minimum of 600 feet upwind until further advised by the incident commander).

        (3) Evacuate areas around suspected device.

            (a) Do not touch or disturb devices.

            (b) Be alert for additional, secondary devices.

        (4) Do not enter enclosed areas (i.e., buildings).

        (5) Contamination.

            (a) Avoid liquid contamination.

            (b) Decontaminate immediately if exposed to liquid contamination.

            (c) Officers inside of the hazard area should exit and be decontaminated as soon as possible.

b. Make notifications.

    (1) Notify dispatch; request supervisor and additional patrols.

    (2) Report critical information.

        (a) Type of injuries and symptoms.

        (b) Estimated number of victims.

        (c) Size or boundaries of the affected area. May be estimated based on the distribution of victims.

        (d) Safe access routes.

        (e) Witness information.

        (f) Warning: Avoid use of cell phones and radios within 300 feet of any suspected device.

    (3) In a suspected CBRNE event, at a minimum the following agencies shall be contacted:

        (a) NCIS.

        (b) Fire department.

        (c) HAZMAT.

        (d) Medical/EMS.

        (e) EOD.

        (f) Other emergency responders (public works, family services, etc.).

        (g) PAO.

2. Wear protective equipment, but know its limitations.

    a. Level C provides adequate protection on the perimeter where live victims exist.

    b. Full-face respirators are necessary for respiratory protection.

    c. Chemical-protective gloves.

    d. Chemical-protective suit.

    e. Foot covers.

N.3 SCENE CONTROL

1. Perimeter definition.

    a. Coordinate with fire department.

    b. Isolate the area.

    c. Corral victims into casualty collection areas away from the area where the agent was released.

    d. Control ingress and egress from the scene.

2. Staging area determination.

    a. Identify an area for responding NSF.

    b. Locate site upwind of the incident scene.

    c. Inform responding units and dispatch of staging area location.

    d. Identify a staging area officer to control the flow of responding officers.

3. Outer-perimeter responsibilities.

    a. The outer perimeter of a chemical agent incident can be expected to be fairly large until the extent of the airborne contamination can be determined. The outer perimeter should be outside of the protective action zone. For determination of the protective action zone, responding NSF and dispatch should confer with fire department and HAZMAT officials.

    b. As monitoring information becomes available, coordinate with the fire department to reduce the size of the outer perimeter.

    c. The primary mission of LE on the outer perimeter will be traffic control.

    d. Additional considerations:

        (1) PPE should be immediately available at the post or in the patrol vehicle.

        (2) The release of additional agent through secondary devices may result in contamination outside of the initial protective action zone.

        (3) Residual contamination may exist on the clothes of victims in and fleeing from the area. Contact with victims, particularly any exhibiting agent signs/symptoms, should be avoided.

        (4) NSF may be required to physically detain someone fleeing the area that may be contaminated.

        (5) Dispatch and NSF should monitor patrols for distress and consider the rotation of officers due to the restrictions of wearing PPE (heat buildup, dehydration, etc.).

4. Inner-perimeter responsibilities.

    a. The inner perimeter is defined by the Emergency Response Guide as the initial isolation zone and the contamination reduction area (commonly referred to as the warm zone and decontamination corridor, respectively).

    b. The area inside of the initial isolation zone is considered contaminated and should be entered only by personnel wearing the appropriate PPE based on the type and level of contamination present.

    c. NSF performing duties on the perimeter of the initial isolation zone should wear appropriate PPE.

    d. The presence of live victims inside of the initial isolation zone does not indicate that the area is free of contamination. Not all agents act immediately, and a physical contact hazard may still exist although the vapor hazard has dissipated.

    e. NSF will most likely perform the following actions on the inner perimeter:

(1) Conducting crowd control.

(2) Assisting fire department and EMS in segregating victims based on priorities for treatment and decontamination.

(3) Securing the initial isolation zone.

(4) Maintaining entry control to/from the area to ensure no one enters without proper PPE.

(5) Securing the decontamination corridor.

(6) Detaining suspects.

(7) Controlling contraband found on victims.

(8) Controlling and conducting initial interview of persons identified as having information regarding the incident.

f. Additional considerations:

(1) Officers should be rotated due to the restrictions of wearing PPE (heat buildup, dehydration, etc.).

(2) NSF must process through decontamination prior to release from the scene.

(3) Long-term security should be posted over the immediate incident site.

5. Responding to an officer down in a CBRNE environment.

a. Notify dispatch, request EMS, and maintain a safe distance and be upwind of site to secure the scene.

b. Attempt to determine the nature of the problem (chemical agent–related, trauma, etc.).

c. Assume chemical agent is present unless able to verify otherwise (e.g., direct communications with the officer).

d. Consider the possibility of additional devices in the area (explosive or chemical).

e. Consider that the agent concentration at the location may be higher than elsewhere.

f. NSF on scene shall request through dispatch that Fire and Emergency Services provide support to assess any CBRN event. NSF shall enter the area only after gaining HAZMAT technician approval and conduct rescue using appropriate PPE levels based on the following factors:

(1) Officer down is inside an enclosed area (i.e., building) and is unable to communicate.

(2) There are no live casualties in the immediate vicinity.

(3) It cannot be determined whether there are any live victims in the area.

N.4 DECONTAMINATION CONSIDERATIONS FOR LAW ENFORCEMENT

1. LE is faced with several key functions in the decontamination process. These are listed below, along with critical self-protection measures.

a. General considerations.

(1) Responding NSF must be proficiency-exercised frequently and be proficient in decontamination procedures prior to a chemical incident occurring.

(2) Responding NSF should be rotated frequently due to the restrictions of wearing PPE (heat buildup, dehydration, etc.).

(3) NSF must be processed through decontamination prior to being released from the scene.

b. Security of personal belongings.

(1) Ideally, everything inside the initial isolation zone (warm zone) stays inside until it is decontaminated and monitored (clothing, personal belongings, etc.).

(2) Collection of personal belongings is expected to be a shared task between the fire department performing decontamination and LE.

(3) Citizens may refuse to disrobe and/or leave personal items (wallets, purses, cell phones, keys, etc.) behind. Allowing citizens to bag small, high-value items may be more prudent than collecting them by force.

(4) Segregate, bag, and label items by individual for future reference.

(5) Evidence pertaining to the incident may be among the collected items.

(6) Safety dictates collection of the items until they are determined to be free of contamination.

(7) The rules governing search and seizure of personal property still apply.

(8) Secondary devices targeting first responders in the decontamination area may be among the collected items. MWD should be used to sweep and clear the decontamination area. Be cognizant of items such as backpacks and briefcases.

c. Security of law enforcement equipment.

(1) NSF who were inside the initial isolation zone must go through decontamination.

(2) A separate decontamination area for first responders should be established in coordination with the fire department.

(3) Equipment and personal belongings should be removed and left at the decontamination area until cleared of contamination.

(4) Sensitive LE equipment (weapons, badges, radios, etc.) should be secured within the decontamination area. Security containers (lock boxes) should be used to secure sensitive LE equipment. Security containers should be under the control of NSF at all times.

(5) NSF should reequip with clean items necessary to continue their mission and/or return to duty.

(6) Monitoring of items for completion of decontamination should be coordinated with HAZMAT personnel.

2. Incident investigation.

a. NSF are responsible for crime scene preservation, in support of NCIS.

b. NSF shall not operate in the hot zone or employ level A or B PPE unless properly trained, certified, equipped, and exercised as an evidence recovery team (ERT). ERTs may be employed in specific remote locations, OCONUS, or U.S. territories and possessions due to the inability to effect timely evidence recovery by federal authorities. Written approval to establish ERTs must be provided by fleet commanders with concurrence from CNIC as the resource sponsor. NCIS will normally collect and process evidence, ensure chain of custody is maintained, and interview witnesses for these events.

c. Security of agent samples.

   (1) HAZMAT teams operating in the attack area (hot zone) can be expected to collect samples for the purpose of identifying the agent. These samples may be considered part of the evidence from the scene and should be controlled by LE accordingly.

   (2) Chain of custody should be started as soon as possible.

   (3) Samples used for the initial hazard analysis should be processed expeditiously by HAZMAT decontamination teams. Samples can be placed in secondary containers or double-bagged to reduce the contact hazard with the sample container.

   (4) Samples to be used for formal evidence should remain in the decontamination area under control of LE until the outside of the containers is decontaminated and determined to be free of contamination.

   (5) At no time will untrained NSF personnel conduct cleanup operations of biohazardous material.

d. Interview of witnesses.

   (1) Focus on those who were closest to the point of dissemination first. Most likely they will be the more severely injured and ill.

   (2) Questioning of potential witnesses should be done as quickly as possible; however, do not delay medical treatment and transport for investigative purposes. If witnesses have not undergone decontamination, officers conducting interviews will need to wear proper PPE.

   (3) Consider that the suspect may be among the victims/witnesses.

   (4) In addition to normal crime scene investigation questions, questions should focus on the key indicators of a chemical agent incident (spray devices, PPE, mists, etc.).

   (5) Identify victims who left the scene on their own and reported for medical treatment at hospitals, clinics, and private doctors' offices, in coordination with the health department.

   (6) Advise dispatch of those victims who left the scene on their own. Consider establishing an incident hotline/tip line.

   (7) Coordinate with the fire department to prioritize decontamination of persons identified as having information regarding the incident ahead of routine victims.

   (8) Provide investigators at casualty collection/treatment facilities.

   (9) Consider that devices may have captured recordings/photos of the event and that they may be located in the personal belongings collected during decontamination processing.

N.5 BIOHAZARD PROTECTIVE EQUIPMENT

NSF members will have biohazard protective equipment immediately available to meet foreseeable mission requirements (see figure N-1).

NOMENCLATURE	DESCRIPTION	BASIS OF ISSUE
Basic Biohazard Kit	1 - Full-length impervious jumpsuit with hood and attached shoe covers 1 - Fluid shield procedure mask with splash-guard visor 1 - Pair of latex medical gloves 1 - Biohazard waste bag 1 - Antimicrobial towelette	2 per emergency vehicle, one per ECP, plus 1 for every 10 NSF members for reserve stock and training
Disposable Cardiopulmonary Resuscitation (CPR) Mask	CPR mask with one-way valve and built-in mouthpiece with hard case	2 per vehicle
Gloves	8 mil examination gloves, latex, and/or vinyl	2 pair for each NSF member, with reserve stock to meet mission requirements

Figure N-1. Minimum Required Equipment for Navy Security Force First Responders

INTENTIONALLY BLANK

APPENDIX O

Physical Security

O.1 SECURITY EDUCATION PROGRAM

1. Security is a responsibility of every employee of DON—military, civilian, and contractor. NSF cannot accomplish their mission without the active interest and support of each member of the command.

2. COs should ensure that a security education program—including crime prevention, protection, and other required security training programs, including counterterrorism briefs, operational security, computer security, information security, and personnel security—is provided for all hands IAW the guiding directives for those programs. The goal of the program should be to keep all personnel vigilant and concerned. Security education program requirements shall include ensuring that all assigned personnel (military, civilian, and contractor) recognize and understand their responsibilities and roles and the pertinent aspects of protection.

3. All personnel shall receive initial security instruction to ensure that they understand the need for security, to help them be proficient in the security procedures applicable to the performance of their duties, and to help them understand the possible consequences of security lapses or breakdowns.

4. Refresher training shall be given to the extent necessary to ensure personnel remain mindful of and proficient in meeting their security responsibilities or as required by the guiding instructions.

5. For all personnel attending required training, training records shall be maintained in either department/division training records or Office of Personnel Management files.

O.2 SECURITY CHECKS

Each installation should establish a system for the daily after-hour checks of restricted areas, facilities, containers, and barrier or building ingress and egress points to detect any deficiencies or violations of security standards. Records of security violations detected by NSF personnel shall be maintained for a period of three years. The SO must follow up each deficiency or violation and keep a record for a period of three years of all actions taken (structural, security, disciplinary, administrative, etc.) by the responsible department to resolve the present deficiency or violation and to prevent recurrence.

O.3 SECURITY INSPECTIONS

1. SOs should ensure that security inspections by trained and designated PS specialists are conducted for all activities possessing restricted areas and facilities and mission-critical areas designated in writing by the CO. Security inspections are conducted at AA&E storage facilities (including RFI storage areas) and other locations aboard Navy installations, activities, and facilities as directed by the commander. Some units/facilities may be exempt from inspection due to their mission. These units/facilities shall be inspected under the guidance of regulations and directives unique to those activities.

2. SOs shall be responsible for inspections of all designated units/facilities within their AO. The security inspection shall provide a recorded assessment of the PS procedures and measures that the unit or activity inspected has implemented to protect its assets. Security inspections shall be maintained on file by the NSF unit and the unit or activity inspected for a period of three years and be available for review by the Navy region and Echelon 2/naval component commands and the Navy Inspector General.

3. Security inspections will be conducted:

   a. When a critical asset is designated or a unit or facility is activated.

   b. When no record exists of a previous security inspection for designated areas, units, or facilities.

   c. When there is a change in the unit or activity that may have an impact on existing AT plans or there is a pattern of criminal activity.

   d. At least once every two years.

   e. When the commander determines that a greater frequency is required.

   f. Inspectors will use DD Form 2637 Physical Security Evaluation Guide (http://www.dtic.mil/whs/directives/infomgt/forms/forminfo/forminfopage2019.html) and DD Form 2638 Waterside Security Evaluation Guide (http://www.dtic.mil/whs/directives/infomgt/forms/forminfo/forminfopage2018.html) as appropriate when conducting security inspections.

4. Security inspectors will be granted access to Navy units, activities, records, and information on a need-to-know basis, consistent with the inspector's clearance for access to classified defense information and provisions of applicable DOD and Navy policy.

5. The inspection should be conducted from the outside to the inside of the unit/facility or area, with regard to the following:

   a. Observation during all hours.

   b. Interviews of managerial and operational personnel.

   c. Observation of and discussion with security forces and personnel (should be inspected so as not to disrupt the mission).

   d. Assessment of security force training.

   e. Evaluation of entry and movement control.

   f. Evaluation of communications systems used by the NSF. (NSF should have two reliable and efficient means of communication, one of which must be a two-way radio.)

   g. Inspections should be conducted according to regulations appropriate for the facility.

6. Recommendations shall be made according to regulations, not personal opinion. Written reports should be forwarded through channels in a timely manner according to the SO's established procedures. Copies of security inspection reports shall be provided to:

   a. Commander of the unit or organization inspected.

   b. Commander of the next higher echelon in the inspected unit's chain of command.

   c. SO of the department conducting the inspection.

   d. Regional commander.

   e. Installation ATWG.

7. COs shall provide a report of corrective actions taken, to the REGCOM and to the activity's next senior in the chain of command. The SO staff shall review this report and provide comments to the RSO and chain of command.

8. Findings noted on security inspection reports that are beyond the capabilities of the REGCOM to correct because of a lack of resources shall be forwarded to CNIC with recommendations and requests for resource assistance. Refer to Appendix U of this NTTP for waiver and exception procedures.

9. Compensatory measures within available resources shall be placed in effect pending completion of work orders or fulfillment of other requests for resource assistance.

10. If a vulnerability assessment is conducted that meets the criteria described herein for a designated facility (see paragraph O.4), the requirement for the security inspection is satisfied. Documentation to this effect shall be identified within security inspection records.

11. Not all Navy assets at all installations require the same degree of protection. Protection of assets must be based on a realistic criticality assessment of the risks associated with the criminal and terrorist threats likely to be directed at the assets in their actual locations. Performing this assessment or risk analyses for assets allows commanders to establish asset protection appropriate for their value and the likelihood of an attempt to compromise them. This assessment allows the commander to prioritize assets so that physical security resources can be applied in the most efficient and cost effective manner possible.

O.4 VULNERABILITY ASSESSMENTS

REGCOMs should ensure that vulnerability assessments of housing areas, facilities, and/or activities at locations and command levels identified as installations within their purview are conducted by either the NCIS security training assistance and assessment team (STAAT) or Defense Threat Reduction Agency (DTRA). The STAAT conducts the CNO Integrated Vulnerability Assessment (CNOIVA), while DTRA conducts the Joint Staff Integrated Vulnerability Assessment (JSIVA) every three years for commands of 300 personnel; any facility bearing responsibility for emergency response and physical security plans and programs; and any facility possessing authority to interact with local nonmilitary or host-nation agencies or having agreements with other agencies or host-nation agencies to procure these services IAW OPNAVINST F3300.53 (series), Navy Antiterrorism (AT) Program. JSIVAs and CNOIVAs are vulnerability-based evaluations of an installation's ability to deter and/or respond to a terrorist incident. A vulnerability-based assessment considers both the current threat and the capabilities that may be employed by both transnational and local terrorist organizations, in terms of both their mobility and the types of weapons historically employed.

Per OPNAVINST F3300.56, Navy Antiterrorism (AT) Strategic Plan, commanders shall conduct an annual vulnerability assessment of all facilities, installations, and operating areas within their AO. These local assessments must include all activities and elements residing as tenants on installations or geographically separated but under the tactical control of the local commander for AT. The requirement for local assessments is satisfied if a higher headquarters vulnerability assessment is conducted at that location in that calendar year.

To receive possible funding to eliminate/mitigate the noted vulnerability, all vulnerabilities identified through an integrated vulnerability assessment or self-assessment must be reflected in the installation core vulnerabilities assessment management program account.

O.5 THREAT ASSESSMENTS

All NCIS components should maintain close and effective liaison with local, state, and federal LE and intelligence agencies and disseminate, by the most effective means, threat information potentially affecting the security of a particular military installation and/or designated facilities on or off base. If a command detects or perceives threat information, the servicing NCIS component should be promptly notified. In installation security operations, a threat analysis is a continual process of compiling and examining all available information concerning potential activities by criminal or terrorist groups that could target an installation, facility, or unit. A comprehensive threat analysis will review the factors of a criminal or terrorist group's existence, capability, intentions, history, and targeting, as well as the security environment within which the installation operates. Threat analysis is an essential step in identifying probability of criminal activity or terrorist attacks and results in a threat assessment. Comprehensive threat assessments must be performed annually or within 30 days of a ship visit. Additionally, NCIS will provide, upon request, an area threat assessment through the servicing NCIS office.

O.6 RISK MANAGEMENT

COs will establish a risk management process to identify, assess, and control risks arising from criminal activities, including terrorism, and to assist in planning and conducting force protection. The risk management process should be embedded into unit operations, culture, organization, systems, and individual behaviors.

O.7 PHYSICAL SECURITY OF ACTIVITIES NOT LOCATED ABOARD NAVY INSTALLATIONS

At all Navy activities not located aboard a Navy installation, the commander shall establish a PS program and appoint a physical security officer in writing. Activities located aboard other DOD Service/agency sites will coordinate PS requirements with the host. Navy organizations should establish interservice support agreements (ISSAs)/MOUs/MOAs with the host Service or nation. Topics that should be addressed in these agreements include establishment and identification of property boundaries, IDS monitoring, available response forces, use of force (including deadly force) training and issues, PS support, etc. Commanders shall coordinate all such agreements through higher headquarters, including the SJA/Office of Counsel/legal offices.

O.8 COORDINATION

To provide for efficient coverage of security needs without wasteful duplication:

1. PS of separate activities and installations shall be coordinated with other military activities/installations in the immediate geographic region or area and with local civilian LE agencies or host-government representatives. Opportunities to partner or share special capabilities among regional users shall be fully explored and documented to ensure economy of effort.

2. Within the physical confines of the installation, the CO shall coordinate PS measures employed by tenant commands.

3. The PS of all AA&E and other HAZMAT held by tenant activities shall be closely coordinated with the SO.

4. All planning that may result in the physical relocation of an organizational element, physical changes to a facility, or a realignment of functions shall include the SO from the outset, to ensure that security considerations are included during initial planning.

O.9 PHYSICAL SECURITY SURVEYS

COs should ensure that PS surveys of their installations are conducted annually and the results addressed as a part of their command review and assessment program. Survey results are a local management tool and not normally disseminated up the chain of command. Surveys will serve to update the command on what needs protecting, what security measures are in effect, what needs improvement, and to provide a basis for determining security priorities. The Physical Security Evaluation Guide (DD Form 2637) is used to document all physical surveys. This form provides a comprehensive template for the conduct of physical surveys. The Waterside Security Evaluation Guide (DD Form 2638) should be used for waterside physical security surveys.

Surveys shall include a complete study and analysis of the activity's property and operation, as well as the PS measures in effect. PS, ESSs, AECSs, IDSs, locks, lighting, barriers, and procedural (access control, lock and key control, property accountability, and training) systems shall be examined. Store and protect PS surveys IAW the information reported (e.g., specific vulnerabilities require a classification dependent on the type of vulnerability or based on the supporting instructions).

In addition to all normally inhabited facilities, surveys shall be conducted at the following facilities.

1. AA&E storage facilities IAW OPNAVINST 5530.13 (series), Department of the Navy Physical Security Instruction for Conventional Arms, Ammunition, and Explosives (AA&E)

2. Exchange and commissary facilities (in coordination with the Navy Exchange Service Command (NEXCOM) and Defense Commissary Agency (DeCA) PS specialists)

3. Storage facilities containing sensitive and/or high-value materials

4. Activities possessing restricted areas and facilities and mission-critical areas designated in writing by the commander

5. The office, headquarters, or residence (if applicable) of the installation command staff and any local or regional senior leadership that regularly works or resides on the installation.

Other Navy activities not located on a Navy installation may request NSF unit PS surveys of their facilities. The RSO will coordinate such requests.

O.10 BARRIERS AND OPENINGS

O.10.1 Barriers and Openings

The SO shall ensure that sufficient barriers are in place to control, deny, impede, delay, and discourage access by unauthorized persons. In determining the type of barriers required, the following shall be considered:

1. Physical barriers (natural or man-made) shall be established or positioned along the designated perimeter of all restricted areas. The barrier or combination of barriers used must afford a minimally acceptable equal degree of continuous protection along the entire perimeter of the restricted area.

2. In establishing any perimeter barrier, consideration must be given to providing emergency entrances and exits in case of fire or explosive hazard. However, openings shall be kept to a minimum consistent with the efficient and safe operation of the facility and with present FPCON, and to minimize the extent of resources required for security.

3. Either access through openings must be controlled, or the openings must be secured against surreptitious entry. Sewers, air intakes, exhaust tunnels, and other utility openings that penetrate the perimeter or restricted area barrier and have a cross-section area of 96 inches or greater shall be protected by securely fastened bars, grills, locked manhole covers, or other equivalent means that provide security commensurate with that of the perimeter or restricted area barrier. Bar grills across culverts, sewers, storm sewers, etc., are a hazard when susceptible to clogging. This hazard must be considered during construction planning. All drains/sewers shall be designed to permit rapid clearing or removal of grating when required. Removable grates located on the perimeter or restricted area shall be locked.

4. If the perimeter barrier is designed to protect against forced entry, then any openings in the barrier must provide protection against forced entry.

5. Vehicle barriers such as crash barriers, obstacles, or reinforced drop arms and/or chain-link gates at uncontrolled avenues of approach will be used where necessary to prevent unauthorized vehicle access.

6. Requirements for standoff distances can be found in UFC 4-010-01, DOD Minimum Antiterrorism Standards for Buildings.

O.10.2 Fences

Navy installations and activities shall use standards and specifications identified in UFC 04-022-03, Security Engineering: Fences, Gates, and Guard Facilities, when using chain-link fencing, gates, and accessories as protective barriers.

O.10.3 Walls

Walls may be used as barriers in lieu of fences. The protection afforded by walls shall be equivalent to that provided by chain-link fencing. Walls, floors, and roofs of buildings may also serve as perimeter barriers.

O.10.4 Temporary Barriers

Additional security forces, patrols, and other temporary security measures will be used where the temporary nature of a restricted area does not justify the construction of permanent perimeter barriers.

O.10.5 Clear Zones

1. Unobstructed areas or clear zones shall be maintained on both sides of the restricted area fences. Correspondingly, where exterior walls of buildings form part of restricted area barriers, an unobstructed area or clear zone shall be maintained on the exterior side of the building wall. Vegetation or topographical features that must be retained in clear zones for erosion control or for legal reasons shall be trimmed or pruned to eliminate concealment of a person lying prone on the ground. Additionally, the vegetation should not be more than 8 inches in height and should be checked by security patrols at irregular intervals.

2. An inside clear zone shall be at least 30 feet. Where possible, a larger clear zone should be provided to preclude or minimize damage from incendiaries or bombs.

3. The outside clear zone shall be 20 feet or greater between the perimeter barrier and any exterior structures or obstruction to visibility.

4. Obstacles may exist within exterior and interior clear zones if they offer no aid to circumvention of the perimeter barrier and do not provide concealment (or provide a plausible reason to appear innocently loitering) to an intruder.

5. Alternatives to extending the clear zone where fences already exist may include increasing the height of the perimeter fence, extending outriggers, or installing double outriggers.

6. Inspections of clear zones should be incorporated with inspections of perimeter barriers to ensure an unrestricted view of the barrier and adjacent ground.

O.10.6 Inspection of Barriers

Security force personnel should check restricted area perimeter barriers at least weekly for defects that would facilitate unauthorized entry and report such defects to supervisory personnel. Installations will establish procedures to document that area perimeter barriers have been inspected for defects.

Personnel must be alert to the following:

1. Damaged areas

2. Deterioration

3. Erosion of soil (intent here applies mostly to instances where a fence is used as the perimeter barrier)

4. Growth in the clear zones that would afford cover for possible intruders and concurrently hinder effectiveness of any protective lighting, assessment systems, etc.

5. Obstructions that would afford concealment or aid entry or exit for an intruder or provide a plausible excuse to openly loiter without need for hiding (e.g., a bus stop next to the fence line)

6. Signs of illegal or improper intrusion or attempted intrusion.

O.11 PROTECTIVE LIGHTING

O.11.1 Lighting Principles

SOs shall apply the following principles when protective lighting is installed and used:

1. Provide adequate illumination to discourage or detect attempts to enter restricted areas and to reveal the presence of unauthorized persons within such areas.

2. Avoid glare that handicaps security force personnel. Carefully weigh security requirements with steps necessary to avoid glare that is objectionable to air, rail, highway, or navigable water traffic or to occupants of adjacent properties.

3. Locate light sources so that illumination is directed toward likely avenues of approach and provides relative darkness for patrol roads, paths, and posts.

4. Lighting at ECP shall be directed toward traffic entering the installation. The light should not be positioned in a direction that will illuminate security personnel. Commands must strive to keep the security force in the shadows as much as possible to minimize exposure. ECP lighting should have a backup generator or an uninterruptible power source system in the event of an installation-wide power loss.

5. Adequate lighting should be installed to protect all avenues of approaches to an installation.

6. Illuminate shadowed areas caused by structures within or adjacent to restricted areas.

7. Provide light distribution that is overlapping and eliminates darkened spaces.

8. Select equipment that is designed to resist the effects of environmental conditions.

9. Provide maximum protection against intentional damage to all components of the system.

10. Meet requirements of blackout and coastal dim-out areas.

11. Consider future requirements of CCTV and recognition factors in selecting the type of lighting to be installed. Where color recognition will be a factor, full-spectrum (high-pressure sodium vapor, metal halide with high-pressure sodium, etc.) lighting vice single-color should be used. Additionally, for CCTV:

    a. Choose lights that illuminate the ground or water but not the air above. Consider light positioning (low or multilevel fixtures) or the intensity of the lighting system (light to 50 fc in lieu of 1/2 fc) to improve capability in fog and rain.

    b. Consider requirement for on-demand infrared lighting.

O.11.2 Illumination of Assets

In coordination with the RSO, the SO will decide what areas or assets to illuminate and what lighting systems to use using the following considerations:

1. Relative cost value of items being illuminated

2. Significance of the items being protected in relation to the activity mission and its role in the overall national defense structure

3. Availability of security forces to patrol and observe illuminated areas

4. Availability of clear zones

5. Availability of fiscal resources (procurement, installation, and maintenance costs)

6. Energy conservation.

O.11.3 Restricted Areas

Restricted areas provided with exterior protective lighting should have an emergency power source located within the restricted area. Emergency power systems shall be tested at least quarterly. Installations will establish procedures to document that emergency power systems have been tested.

O.11.4 Wiring Systems

Both multiple and series circuits may be used depending on the type of luminary used and other design features of the system. The circuits in protective lighting systems should be arranged so that the failure of any one lamp will not darken a section of a critical or vulnerable area. The restricted area protective lighting system should be independent of other lighting systems so that normal interruptions caused by overloads, industrial accidents, and building or brush fires will not interrupt the protective system.

O.11.5 Controls and Switches

Controls and switches for restricted-area protective lighting systems should be inside the protected area and otherwise secured. High-impact plastic shields should be installed over lights to prevent destruction.

O.11.6 Power Sources

Power supplies related to lighting systems protecting Level A and B assets should be routed to the installation separately from other utility services.

O.12 SECURITY OF NAVY FACILITIES

O.12.1 New Construction and Facility Modifications

All new construction shall comply with COCOM requirements and the requirements of UFC 4-010-01, Department of Defense Minimum Antiterrorism Standards for Buildings, as well as pertinent facility modifications to existing buildings, facilities, sites, etc. The ISO or designated representative shall review plans for new construction and facility modifications during the design process and various review phases to ensure that PS, loss prevention, AT, and force protection measures are adequately incorporated. The operational Echelon 2 commander will address issues that cannot be resolved at the local level because of lack of necessary funding or other reasons outside the control of the local command (e.g., appropriate and adequate clear zones).

O.12.2 Navy Military Construction Projects

Submit proposals for Navy military construction (MILCON) projects (major construction estimated to cost more than $400,000) through the respective Navy region, then through the Commander, Naval Facilities Engineering Command (COMNAVFACENGCOM), and finally through CNIC for endorsed approval. The Naval Facility Engineering Command (NAVFAC) has responsibility for a review of PS construction projects, to include ensuring that this instruction is adhered to or a waiver is submitted via the same chain. In fulfilling this responsibility, NAVFAC will make certain that protective design measures and equipment reliability and maintenance have been considered.

O.12.3 Security of Leased Facilities

Responsible commands should:

1. Address PS in all lease agreements.

2. Liaison with appropriate authorities (e.g., GSA, building administrators, leasers, etc.) will delineate specific security responsibilities among the concerned parties regarding measures that are necessary for the protection of lives and property and that are tailored to the individual characteristics of the leased space.

3. Identify PS standards that cannot be met, either temporarily or permanently, and submit waiver or exception requests, as appropriate, per Appendix U of this NTTP. Compensatory security measures implemented or planned must be identified in all such waiver or exception requests.

O.12.4 Activity Upgrade Requirements

Any new construction, upgrade, or modification to existing facilities should not reduce the security posture or create a new vulnerability (as would be identified in security surveys, inspections, or vulnerability assessments). A plan of action and milestones shall be developed to correct identified deficiencies. Deficiencies that are not correctable within 12 months shall be covered by an approved waiver or exception pending completion of the required upgrade effort. Compensatory security measures are required.

O.13 FLIGHTLINE AND AIRCRAFT SECURITY

O.13.1 Flightline and Aircraft Security Planning

In planning for flightline and aircraft security, commanders should consider:

1. The security provided by the host installation or commercial airfield (e.g., perimeter security, ECP measures, security patrols).

2. Whether the flightline or aircraft parking area is adequately fenced, lighted, and posted with signs and the procurement, installation, and maintenance of physical barriers, fencing, lighting, etc.

3. Security or LE patrol coverage.

4. Security consciousness of personnel working within the flightline.

5. IDS or other PS procedures, devices, and equipment used to protect the flightline.

6. Coordination with the tenant commands for the return of deployed units' personnel and equipment.

7. Establishment of aircraft parking plans.

8. Designation of the flightline and aircraft parking area restricted areas.

9. Designation of vehicle parking areas.

10. Policy on the use of taxiways and runways by security/safety personnel.

11. Approval of ECPs for the flightline restricted area.

12. Personnel and vehicle access control for the flightline restricted area.

13. Flightline security representation on the ATWG.

14. Incorporation of flightline security issues into the installation AT plan(s).

15. Potential threat locations for man-portable air defense system (MANPADS) surface-to-air missile (SAM) systems.

16. Increased NSF actions to reduce MANPADS threat. Doctrinal information on security against MANPADS threats is provided in DOD O-2000.12H, Department of Defense Antiterrorism Handbook (FOUO). SOs should also coordinate these actions with local LE agencies and solicit their assistance in defending against this threat.

17. Vulnerabilities created by any unprotected openings from utilities and natural topography (electrical, gas, water, drainage ditches) entering the airfield.

18. Use of intrusion sensors capable of sensing multiple indicators of an intrusion (sensed phenomenologies) to assist in classifying and identifying intrusion alarms.

19. Potential for threats to necessitate security force beyond-the-perimeter surveillance and response actions.

O.13.2 Installation Security Officer Responsibilities

The SO is responsible for planning, organizing, and directing security of the flightline and visiting aircraft, including:

1. Ensuring that the database that allows flightline access is integrated into and supports the overall installation AT plan(s).

2. Issuing restricted area access badges/common access cards.

3. Publishing RRP for flightline security operations per the provisions of this NTTP.

4. Providing flightline restricted areas protection by motorized patrols and security response teams, as well as coordinating with tenant commands for augmentation forces when required.

5. Conducting annual PS surveys on flightline restricted areas.

6. Making recommendations concerning the location and types of barriers to be employed in and around flightline restricted areas.

7. Reviewing plans for the acquisition and construction of barriers by tenant organizations (e.g., work requests, budget items, purchase orders, etc.) to ensure compatibility with security systems aboard the installation.

8. Approving deployment and removal of barriers.

9. Ensuring that NSF personnel inspect barriers at least monthly for good repair and operability. Problems will be documented and reported to the appropriate activity for corrective action.

10. Ensuring NSF personnel are educated on the MANPADS threat, including component recognition, areas of vulnerability, and reaction plans.

11. Developing and exercising contingency plans for responding to an incident of a MANPADS threat.

O.13.3 Flightline Security

1. Flightline security is the capability to protect the flightline and its transient aircraft from encroachment and attack. Flightline security duties such as control of entry points, patrol, surveillance, and emergency response shall be performed by NSF under the operational control of the SO.

2. Personnel augmenting the flightline security program (ASF, squadron personnel) shall be screened and trained by the security training division prior to assignment as part of the security force. The SO has final approval of temporary personnel assignment.

3. ECPs should be equipped with gates, gatehouses, areas for processing personnel and vehicles into the flightline restricted area, communications equipment (two-way radio and telephone), and security lighting.

4. Entry to the flightline restricted area will be authorized only at designated ECPs (pedestrian or vehicle) that will be manned by armed sentries, access control personnel, or an AECS IAW with procedures for the protection of Level Two or Three restricted areas, as detailed in Chapter 2 of OPNAVINST 5530.14 (series), Navy Physical Security and Law Enforcement Program.

5. Personnel not authorized access shall be escorted to the unit by designated personnel from the unit being visited. Dependents may be admitted to conduct official business upon presentation of a valid Uniformed Services Identification and Privilege Card (DD 1173), if the sponsoring unit provides an escort. Visitors shall not be granted unescorted access to the flightline restricted area.

6. The CO may waive access control measures during special events (ceremonies, air shows, etc.) provided that additional security measures are taken during the event.

7. Contractors and delivery personnel may be authorized access by the SO on a case-by-case basis. Each company will submit a roster, updated quarterly, that identifies personnel who require flightline access to conduct business.

8. Immediate access shall be granted to all prearranged responses by emergency vehicles responding to locations within the restricted area or according to the AT plan. Emergency vehicles are ambulances, fire trucks, NSF police vehicles, crash trucks, and EOD vehicles. Such vehicles should not be impeded, and security personnel will render assistance as required. NCIS special agents shall be admitted upon presentation of their credentials.

9. Government vehicles authorized within aircraft parking areas shall be clearly marked. The manner of marking will be coordinated with the security unit. After normal operating hours, personnel shall contact dispatch to advise of any requirement to move government vehicles within aircraft parking areas.

10. All flightline restricted areas shall use an integrated barrier system tailored to meet the unique characteristics of the site.

11. Flightlines, hangars, and similar restricted areas should be equipped with security fencing and lighting.

    a. Restricted areas where aviation safety precludes fencing will be protected by IDS.

    b. Temporary barriers may be used during increased FPCON or pending installation of fencing, IDS, or other permanent equipment.

12. Flightline and host installation security patrols shall be briefed on the current local threat assessment, including MANPADS SAM. Appropriate post and patrol procedures should include guidance on recognition of and response to such threats.

O.13.4 Tenant Commands

Tenant commanders at the squadron level and higher will appoint in writing an access control officer who shall:

1. Submit to the SO a roster listing those individuals:

   a. Authorized access to that unit's area

   b. Who require a locally issued restricted area access badge.

2. Submit this roster the first month of each quarter.

3. Report all losses of security badges or command access cards to the SO. Reporting must be done immediately so that the database can be modified to deny access to that particular card.

O.13.5 Privately Owned Vehicle Parking

COs, acting on the advice of the SO, may designate parking areas in and around flightline restricted areas as dictated by the local threat condition and operational necessity. An accountability system shall be devised locally to track all POVs given access. POV parking within flightline restricted areas will not require a waiver or an exception provided that an accurate accountability system is used.

O.13.6 Aircraft Parking Areas

1. COs shall develop aircraft parking plans that include dedicated aircraft parking areas. Plans shall provide for the following:

   a. The number of aircraft parking areas should be limited and consolidated with or located adjacent to other support assets within the flightline restricted area.

   b. When modification of aircraft parking areas is necessary, the SO shall be notified so that PS requirements may be coordinated prior to modification.

   c. Aircraft parking areas should be clearly marked.

   d. When determining location, consideration should be given to proximity to public areas, avenues of approach, and response routes for use by NSF personnel.

2. Each aircraft parking area shall be provided with surveillance.

   a. Under FPCON NORMAL, unit operations and maintenance personnel may fulfill the requirement for surveillance. Aviation personnel can assist NSF personnel by reporting any condition that threatens security.

   b. The use of electronic surveillance equipment may be used as a means of augmenting and/or reducing the number of NSF required; however, it will not be used as the sole means of surveillance.

O.13.7 Transient or Deployed Aircraft

1. COs shall provide a secure area for transient aircraft aboard Navy installations.

2. The host installation should make every reasonable effort to provide the same degree of security that the owning Service would provide under the same (transient or deployed) circumstances.

   a. Transient aircraft should be parked in a permanent restricted area with employed IDS.

b. Transient aircraft may be parked in a hangar or encircled with an elevated barrier, such as rope and stanchions, if it is not possible to park the aircraft in an established restricted area with IDS. When hangers are used, the walls will constitute the restricted area boundary.

c. Area lighting of sufficient intensity shall be provided to allow NSF to detect and track intruders.

d. Restricted area signs shall be displayed so personnel approaching the aircraft can see the signs.

e. Circulation control shall be provided to ensure entry is limited to only those persons who have a need to enter.

f. The SO or his representative shall give the aircraft commander a local threat assessment for the duration of ground time.

3. Before operations commence, the command, military Service, or other agency owning the assets should request any special security support from the host installation as far in advance as possible. Economic and logistical considerations dictate that every reasonable effort be made by the host installation to provide the necessary security without resorting to external support from the command owning the asset (aircraft, ship, etc.). The owning command should provide material and personnel for extraordinary security measures (extraordinary security measures are those that require heavy expenditures of funds, equipment, or manpower or unique or unusual technology) to the host installation.

4. Aircraft commanders or their designee shall notify NSF before a visit to the aircraft is allowed to take place. Any change in security priorities based on operational status must be identified to the host installation.

5. Figure O-1 provides minimum security requirements for non-alert aircraft. OPNAVINST 5530.14 (series), Navy Physical Security and Law Enforcement Program, defines each asset level. These requirements apply to operational aircraft on display or located at civilian or foreign airfields. These requirements do not apply to decommissioned museum aircraft. Special or increased requirements for specific operational configuration must be identified in advance (when possible) to host security forces.

a. All priority aircraft require armed response team (ART) support. ARTs may be area patrols not specifically dedicated to the visiting aircraft.

b. Close boundary sentries (CBSs) are NSF posted inside or outside the boundary, to keep the boundary of the restricted area under surveillance.

AIRCRAFT TYPE	ENTRY CONTROL RESPONSIBILITY	ART	CBS	MOTORIZED PATROL
Tactical Aircraft (AV-8, F-14, F/A-18)	Aircrew	Yes		Yes
Airlift Aircraft (C-5, C-9, C-17, C-130, C-141)	Aircrew	Yes		Yes
Strategic Bomber Aircraft (B-52, B-1)	Aircrew	Yes		Yes
Air Refueling Aircraft (KC-10, KC-135)	Aircrew	Yes		Yes
Special Mission Aircraft (E-2, EA-6, EP-3)	Security	Yes	Yes	
Reconnaissance Aircraft (S-3, P-3)	Aircrew	Yes		Yes
Advanced Technology Aircraft	Pilot carries detailed information for divert contingencies			
Other DOD Aircraft	Aircrew	Yes	Yes	Yes

Figure O-1. Security Requirements by Aircraft Type

O.13.8 Off-Base Aircraft Mishaps

1. Initial security for aircraft that crash or are forced to land outside a military installation is the responsibility of the nearest military installation. The owning Service will respond and assume on-site security as soon as practical.

2. The incident commander shall be responsible for coordinating with civilian LE. Such coordination may include:

   a. The methodology for granting access to the accident scene at roadblocks

   b. Location of EOC

   c. Names and locations of senior supervisors (crash/fire rescue, medical, civilian LE, military security, aircraft mishap board, public affairs, Federal Aviation Administration representative)

   d. Safety precautions.

3. In the above emergency situations, NSF must:

   a. Ensure the safety of civilian sightseers.

   b. Prevent tampering with or pilfering from the aircraft.

   c. Preserve the accident scene for later investigation.

   d. Protect classified cargo or aircraft components.

O.13.9 Security of Navy Aircraft at Non-Navy Controlled Airfields

Navy aviation units possess no organic security forces. Trained security personnel composed of U.S. military security units, HN security forces, private security personnel, or a combination of these must provide force protection for expeditionary air forces ashore. Such forces must meet standards approved by the GCC for the airfield in question. NWP 3-10, Maritime Expeditionary Security Operations contains further guidance regarding aircraft security missions.

When planning exercises or operations that involve Navy aircraft, adequate U.S. military security forces or appropriate augmentation must be provided. Mobile NSF are requested through the appropriate numbered fleet commander for such duties. Thoughtful consideration of the combination of local conditions, terrorist threat, and FPCON is required to determine specific augmentation requirements.

Commander, Naval Air Forces (CNAF) will promulgate guidance for naval aircraft transiting in the continental United States (CONUS), Alaska, and Hawaii.

O.14 WATERSIDE AND WATERFRONT SECURITY

O.14.1 General

Regional commanders shall ensure waterways adjacent to afloat assets and supporting infrastructure are under appropriate surveillance and adequately patrolled. In Navy-controlled ports, the host AT plan(s) shall detail the port security posture. In non-Navy ports, post manning and arming are specified in the ship's security plan, which takes into account local LE and/or HN agreements, and are approved by the appropriate higher authority. In private shipyards, Naval Sea Systems Command and applicable supervisor of shipbuilding shall ensure that contractors fulfill security requirements. Doctrinal information pertaining to waterside PS measures can be found in DOD O-2000.12H, Department of Defense Antiterrorism Handbook (FOUO).

O.14.2 Security Requirements

1. Waterfronts (including all piers, wharves, docks, or similar structures to which vessels may be secured) shall be designated as restricted areas.

2. Security planning for waterside and waterfront security should include vehicle patrols, access control sentries, cover sentries, roving foot patrols, stationary posts, and electronic harbor security systems. Planning will address additional measures during heightened FPCONs. Water barriers required to protect home-ported assets shall be placed to achieve sufficient stand-off from units being protected. Regional commanders shall develop plans to increase the FPCON at piers and surrounding waters that berth high-value units. Consistent with operational readiness, every effort should be made to get ships underway during FPCONs CHARLIE and DELTA.

3. NSF personnel shall be supplied from the host installation. When installation assets are not adequate, ASF from tenants and augmenting personnel from ships will be assigned. The host installation shall supervise and have authority over all on-duty security personnel during the watch. The SO should ensure that at least one field supervisor is dedicated to waterside security at all times.

4. NSF shall be equipped with a communications system. Two-way radios used by all personnel providing security for waterside assets should be compatible, including communications with NSF personnel, Navy and Military Sealift Command (MSC) ships' personnel, harbor patrol boats, Marines from amphibious ships, USCG, and HN security, where applicable. Communications planning should define interoperability of shipboard security forces and NSF ashore, including harbor security units and communications with local civilian LE.

5. Barriers shall be available to prevent direct unchallenged access to piers, wharves, or docks when ships are moored. Installation commanders shall ensure that barriers are closed at all times when not opened to facilitate ship movements.

6. Vehicle access to piers, wharves, or docks shall be controlled. Parking during FPCONs NORMAL and ALPHA will be limited to essential government or vetted commercial and approved ship's company vehicles. Parking on piers shall be discontinued at FPCON BRAVO and higher.

7. Response force personnel shall be mobile or have adequate security vehicles immediately available for emergency response situations.

8. Figure O-2 provides a description of the security required to protect categorized Navy assets as identified in OPNAVINST 5530.14 (series), Navy Physical Security and Law Enforcement Program. Threat and vulnerability assessments must also be taken into consideration when visiting a port. Each level incorporates the measures from all the previous levels. The use of water barriers for Priority C assets should be considered on a case-by-case basis after evaluating the threat, vulnerability, and asset, and if co-located with A or B assets.

9. While at anchorage inside installation restricted areas, all ships shall be provided protection.

10. Foreign naval ships visiting U.S. Navy installations shall be provided protection commensurate with ships pierside.

11. Piers where ships are moored shall:

    a. Be protected by a pier ECP sentry controlling access at the foot of the pier or by a single ECP established separately from the brow when a unit is moored. No sentry is required for pedestrian-only ECPs controlled by an AECS when the entire perimeter of the restricted area is adequately secured, unless there is an additional requirement to process visitors.

    b. Be augmented with at least one additional person during normal working days or during other periods of heavy traffic, to support visitor control and maintain traffic flow.

    c. Have pier ECP access gates closed at all times except when opened for vehicle access.

    d. Have cover sentry protection at ECPs during FPCON CHARLIE and DELTA. Commander will determine weapon type and placement, based on the tactical situation and type of asset being protected.

12. The primary mission of the waterborne patrol shall be to deter unauthorized entry into waterside restricted areas, maintain perimeter surveillance, and intercept intruders prior to their approaching Navy ships in port. Refer to NTTP 3-20.6.29M, Tactical Boat Operations for further guidance.

    a. Waterborne patrols are required 24 hours per day, 7 days per week. For installations with Priority A assets, patrols shall be continuous. For installations with Priorities B through D assets, patrols may be random during FPCONs NORMAL and ALPHA. However, NSF patrol craft must be in the water (crew nearby) and ready to get underway immediately. Commanders will determine the frequency of the random patrols at FPCONs NORMAL and ALPHA. FPCONs BRAVO and higher require continuous patrols whenever Priorities A through C assets are present.

    b. Naval weapons stations may be provided waterborne patrols dependent on the number of days combatant ships are pier side.

    c. HSB patrol crew shall at a minimum include a complement of two personnel. The number of crew members assigned will be based on the assets being protected and waterfront area to be patrolled. With the exception of coxswain, the crew must be trained security personnel armed with authorized handguns and with at least one rifle, shotgun, or light/medium/heavy machinegun. Additionally, nonlethal 12-gauge airborne munitions or flares (gun or pencil) shall be readily available. Commands shall ensure the patrol crew is informed of the purpose for different flare colors issued. At FPCON BRAVO a light/medium/heavy machinegun will be mounted for use, with ammunition available. In FPCONs CHARLIE and DELTA the HSB will be issued concussion grenades if the commander determines that a swimmer/diver threat exists. The HSB's coxswain may be unarmed nonsecurity personnel.

PRIORITY	ASSET (Reference OPNAVINST 5530.14 (series))	SECURITY MEASURES (Cumulative from low to high)
A (Highest)	Ballistic missile submarines	Per SECNAVINST S8126.1 series at nonhome ports where cumulative asset in port time exceeds more than 120 days in a calendar year. Use water barriers to stop small-boat threat.
B (High)	Carriers, other submarines, and large-deck amphibious (LHD, LHA, etc.)	Electronic water/waterside security system (CCTV, associated alarms, surface craft or swimmer detection, underwater detection). Use water barrier(s) to prevent direct unchallenged access from small-boat attacks. Submarines and aircraft carriers shall receive an armed escort during ingress and egress.
C (Medium)	Surface combatants, other amphibious, auxiliary, MSC Strategic Sealift Ships (SSS), ammunition ships, mine warfare	Establish security zone with USCG where possible. Use water barrier(s) where appropriate and/or practical. Harbor patrol boat(s) with bullhorn, night vision device, spotlight, marine flares, and lethal and nonlethal weapons. Arrange patrol boat backup support from Harbor Ops, USCG, or other (tenant boat units, small craft from ships).
D (Low)	Patrol coastal, MSC SSS (reduced operational status), pier facilities	Adjacent landside security (patrols, surveillance, pier access control), no special requirement in waterways. Identify restricted-area waterway(s) with buoys and signs.

Figure O-2. Security of Waterfront Assets Matrix in U.S. Navy–Controlled Ports

d. During FPCONs CHARLIE and DELTA waterborne patrols shall be deployed at each slip where an aircraft carrier, large-deck amphibious ship, or strategic-lift MSC ship is berthed. In the absence of sufficient numbers of naval facility HSBs, armed picket boats sourced from the Navy ships in port may be used. These boats should:

(1) Maintain positive communication with shore emergency dispatch, HSBs, pier watch standers, and any NSF deployed ashore to intercept small craft and respond to swimmer sightings.

(2) If possible, be sourced from aircraft carriers or large-deck amphibious ships, when present.

(3) Be manned with armed, trained personnel equipped with buoyant and level IV ballistic protection, when available.

e. Submarines and MSC ships berthed alone should coordinate with the CO to provide picket boat coverage.

f. If a subsurface threat exists based on the standing port integrated vulnerability assessment, current threat assessment, or as provided in the unit AT plan, additional waterborne security assets should be considered.

g. In Navy-controlled ports, the host installation is responsible for implementation of waterborne security. In non-Navy ports, the ISIC specifies the requirements in an approved in-port security plan (ISP). In private shipyards, Naval Sea Systems Command shall coordinate security with Navy and local resource providers.

13. Waterfront fighting positions for nonnuclear weapons security areas shall be established as follows:

a. FPCONs NORMAL/ALPHA and BRAVO: Posted such that overlapping fields of fire are provided to protect HVUs. The location and number of fixed fighting positions will be based on the asset to be protected, geography, and pier layout. This determination shall be documented in the installation AT plan. Fighting positions will be armed with light, medium, or heavy machineguns. Flares shall be available for signal and alert notifications. This post may consist of a single person unless manning a crew-served weapon, in which case the team will consist of two persons.

b. FPCONs CHARLIE and DELTA: Waterfront fighting positions shall be positioned at fixed locations to protect all Priority A, B, and C ships. Concussion grenades shall be made available.

14. For all FPCONs, submarines and aircraft carriers shall receive an armed escort during ingress to and egress from port. Escort shall be at a minimum from/to the point where the submarine or carrier is no longer limited in its ability to maneuver (from/to the submarine surface/dive point if possible). The primary mission of the escort will be to interdict any small boats attempting to enter the naval vessel protective zone.

a. At a minimum, two personnel will man the escort. Arming and training requirements are as for HSB. Light, medium, or heavy machineguns shall be mounted and ready for use with ammunition available.

b. In Navy-controlled ports, the host installation is responsible for implementation of submarine escorts. In non-Navy ports, the ISIC specifies the requirements in approved ISP.

c. In the event no Navy assets are available, submarine escorts may be requested from USCG.

O.15 SECURITY OF MATERIAL

O.15.1 General

This appendix provides security policy for safeguarding controlled inventory items, including drugs, drug abuse items (as identified under 21 CFR 1301.71 through 1301.76 and P.L. 91-513), and precious metals. The following definitions describe sensitive items:

1. Selected sensitive inventory items. Those items security-coded "Q" or "R" in the Defense Integrated Data System that are controlled substances, drug abuse items, or precious metals.

2. Code "Q" items. Drugs or other controlled substances designated as Schedule III, IV, or V items, per 21 CFR 1308.

3. Code "R" items. Precious metals and drugs or other controlled substances designated as Schedule I or II items per 21 CFR 1308.

4. Precious metals. Refined silver, gold, platinum, palladium, iridium, rhodium, osmium, and ruthenium in bar, ingot, granule, liquid, sponge, or wire form.

O.15.2 Policy (IAW OPNAVINST 5530.14 series)

1. Controlled inventory items shall have characteristics so that they can be identified, accounted for, secured, or segregated to ensure their protection and integrity.

2. Special attention shall be paid to the safeguarding of inventory items by judiciously implementing and monitoring PS measures. This will include analysis of loss rates through inventories, reports of surveys, and criminal incident reports to establish whether repetitive losses indicate criminal or negligent activity.

O.15.3 Responsibilities

COs will:

1. Establish PS measures to protect inventory items and reduce the incentive and opportunity for theft.

2. Monitor the effective implementation of security requirements through scheduled inspections and staff oversight visits to affected activities.

3. Ensure that adequate safety and health considerations are incorporated into the construction of a security area for controlled inventory items.

4. Ensure that storage facilities and procedures for operation adequately safeguard controlled inventory items.

O.15.4 Controlled Substances Inventory

Accountability and inventory of controlled substances shall be as prescribed in NAVMEDCOMINST 6710.9, Guidelines for Controlled Substances Inventory.

O.15.5 Security Requirements for "R"-Coded Items at Base and Installation Supply Level or Higher

1. Storage in vaults or strong rooms, as defined in SECNAVINST 5510.36 (series), Department of the Navy (DON) Information Security Program (ISP) Instruction, or 750-pound or heavier GSA-approved security containers. Smaller GSA-approved security containers are authorized but must be securely anchored to the floor or wall. All security containers shall be secured with built-in group one combination locks. Alternatively, they may be stored using any means that provide a degree of security equivalent to any of the preceding.

2. Storage areas or containers should be protected with installed IDS.

O.15.6 Security Requirements for "Q"-Coded Items at Base and Installation Supply Level or Higher

1. The preferred storage for sensitive inventory items coded "Q" is in vaults or strong rooms.

2. Small quantities may be stored in security containers or other means approved for items coded "R."

O.15.7 Security Requirements for "R"- and "Q"-Coded Items Below Base and Installation Level (i.e., Small Unit/Individual Supplies)

1. Storage as described in paragraphs O.15.5 and O.15.6.

2. As an alternative, small stocks may be stored in a 750-pound or heavier GSA-approved security container. Smaller GSA-approved security containers are authorized but must be securely anchored to the floor or wall. Also, any means that provides a degree of security equivalent to any of the preceding may be used. Security containers should also be located within a continuously manned space or be checked by a security force member at least twice per 8-hour shift, barring any reason for the contrary.

O.16 SECURITY OF COMMUNICATIONS SYSTEMS

O.16.1 General

1. The protection provided to communications facilities and systems should be sufficient to ensure continuity of operations for critical users and the facilities they support. This determination is based on strategic importance both to the United States and to its allies and on whether or not each mobile system or facility processes, transmits, or receives telecommunications traffic considered crucial by POTUS and SecDef; the Chairman of the Joint Chiefs of Staff (CJCS); or the Global Combatant Commanders. Commander, Naval Network Warfare Command shall be consulted on this issue.

2. Security for communications systems shall be a part of each installation's security program. Security considerations shall be thoroughly assessed for each communications system.

3. Parent Echelon 2 commands shall review the host installation's implementation of PS measures during inspections, oversight, and staff visits.

4. Access shall be controlled to all communications facilities; only authorized personnel, including employees and authorized exceptions with escorts, shall be allowed to enter. Facilities should be designated and posted as restricted areas per this NTTP.

5. Existing essential structures should be hardened against attacks, including large antenna support legs, antenna horns, operations building, and cable trays. Future construction programs for critical communications facilities should include appropriate hardening of essential structures.

O.16.2 Responsibilities

Echelon 2 commands shall:

1. Identify critical communications facilities and mobile systems within their commands.

2. Ensure that a security plan is developed for each communications facility and mobile system within their command. The plan shall include emergency security actions and procedures for emergency destruction of sensitive equipment and classified information. The plan may be an annex to an existing host installation AT plan; only the applicable parts of the total plan will be distributed to personnel at the facility or mobile system.

3. Arrange for security of off-installation facilities and mobile systems with the closest U.S. military installation. This includes contingency plans for manpower and equipment resources during emergencies. These arrangements may be made by establishing a formal agreement such as an ISSA. Whether the facilities are located on or off the installation or are mobile, installation commanders are responsible for security of communications facilities for which they provide host support.

4. Implement a training program to ensure that assigned personnel understand their day-to-day security responsibilities and are prepared to implement emergency security actions. The training program will include the following:

   a. Security procedures and personal protection skills for assigned personnel

   b. The use of weapons and communications equipment for protecting the facility or mobile system

   c. Awareness of the local terrorist and criminal threat and other activity in the area.

O.16.3 Mobile Communications Systems

A security operational concept or standards will be developed for mobile systems to describe the minimum level of security for the system in the expected operational environment.

O.17 PROTECTION OF BULK PETROLEUM PRODUCTS

O.17.1 General

Commanders of government-owned, government-operated and government-owned, contractor-operated fuel support points; pipeline pumping stations; and piers shall designate and post these installations as restricted areas. This requirement does not apply to locations for issue (and incidental storage) of ground fuels for use in motor vehicles, material-handling equipment, and stationary power and heating equipment. Commanders will determine the means to protect against loss or theft of fuel at these locations.

Access to these facilities shall be controlled, and only authorized personnel will be permitted to enter. Commanders shall determine the means required to enforce access control (e.g., NSF, barriers, and security badges) based on the considerations in this instruction.

O.17.2 Antiterrorism Plan and Liaison

Commanders shall take the following actions to protect their fuel facilities:

1. Establish liaison and coordinate contingency plans and inspection requirements with the nearest U.S. military installation, to provide manpower and equipment resources to the facility in the event of emergencies and increased threat conditions, as needed based on the threat assessment.

2. Establish liaison with supporting LE agencies and HN officials and support agreements, if appropriate.

O.17.3 Security Inspections

1. Navy installations responsible for the security oversight of fuel facilities shall conduct a security inspection of that facility at least once every two years.

2. Inspections should be formal, recorded assessments of crime prevention measures and other PS measures used to protect the facilities from loss, theft, destruction, sabotage, or compromise.

3. Installations shall draft a plan of action and milestones to correct discrepancies found.

O.18 NAVY SECURITY FORCES ARMORY OPERATIONS

1. Security RFI armories shall be operated in compliance with the current edition of OPNAVINST 5530.13 (series), Department of the Navy Physical Security Instruction for Conventional Arms, Ammunition, and Explosives (AA&E). Only security weapons, ammunition, and equipment shall be stored in the security RFI, with the following exceptions.

    a. During weekend and holiday routines, personal weapons may be temporarily stored for safekeeping until the next workday.

    b. Storage of personal weapons on an installation shall be authorized by the CO or designated representative in writing. Nongovernment, privately owned weapons not approved for storage in family housing will be stored in an armory or magazine but not in the same security container or weapons rack with government AA&E.

2. The SO shall ensure compliance with the following criteria for storage, issue, and recovery of security weapons and ammunition:

    a. Only General Services Administration (GSA)–approved storage safes shall be used for ammunition.

    b. Inventories shall be conducted at each change of shift.

    c. Primary and alternate weapons custodian shall be assigned in writing by the SO.

    d. All weapons and ammunition transactions shall be documented on the Memorandum Receipt for Individual Weapons and Accessories (NAVMC 10576).

    e. Upon issue, each NSF member shall relinquish his or her Ordnance Custody Receipt Card (NAVMC 10520).

    f. Weapons shall be receipted for at the beginning of each shift.

    g. Weapons loading and unloading shall be accomplished using safe clearing procedures IAW NTRP 3-07.2.2, Weapons Handling Standard Procedures and Guidelines.

    h. All issued weapon(s) and ammunition shall be returned to the RFI arms room upon the completion of the shift.

    i. The individual NAVMC 10520 shall be returned to NSF members upon recovery of the weapon and ammunition.

3. Unescorted access to the RFI arms room shall be granted only to personnel authorized in writing by the SO.

4. All RFI arms room transactions shall be conducted from behind a locked door.

5. Arms room keys shall always be properly secured when not in the possession of an armed individual.

INTENTIONALLY BLANK

APPENDIX P

Loss Prevention

P.1 GENERAL

The loss prevention program is part of the overall command security program dealing with resources, measures, and tactics devoted to the care and protection of property on an installation. It includes identifying and reporting missing, lost, stolen, or recovered government property, including documents and computer media, and developing trend analyses to plan and implement reactive and proactive loss prevention measures. A loss prevention program shall be carried out at every Navy activity.

P.2 LOSS PREVENTION PROGRAM MEASURES AND OBJECTIVES

1. The loss prevention program is a continuing process of loss analysis to identify trends and patterns of losses. It tries to derive loss patterns and trends based on the types of material lost; geographic location; times and dates; proximity of specific personnel; proximity of doorways, passageways, loading docks and ramps, gates, parking facilities, piers, and other activities adjacent to loss or gain locations; material movement paths; etc.

    a. It helps determine the allocation of resources available IAW the results of analysis of loss and gain trends and patterns for crime prevention.

    b. It guides COs to develop action plans to prevent or reduce opportunities for losses of government property at supply centers, shipyards, shipping and receiving points, ordnance stock points, and other Navy activities.

    c. It prioritizes warehouses, storage buildings, office buildings, and other structures that contain high-value, sensitive, or pilferable property, supplies, or office equipment so that security protection is commensurate with the value and sensitivity of the contents.

2. Loss prevention program support shall be included in support for host/tenant shore activities. Such support should be coordinated and integrated among all Navy activities on a regional basis.

3. Owner-user responsibility is the key to program success.

    a. Personnel should be provided training and reminded of the need for loss prevention and local procedures for preventing property losses, as well as their role for the care and protection of government property.

    b. Loss prevention training and briefings should be conducted at least semiannually.

P.3 REPORTING SUSPICIOUS OR HOSTILE ACTS AND CRIME HAZARDS

1. All personnel should be aware of and familiar with personnel in their work areas to determine whether they are authorized.

2. An emergency number or hotline to the dispatcher should be published to encourage the immediate reporting of all suspicious or hostile acts.

3. LE patrols should conduct regular facility checks for potential crime hazards (e.g., unsecure equipment, unlocked doors, broken windows, burned-out light bulbs, etc.). If discovered, the owning CO, unit, and facility custodian must be notified so they may take steps to secure the resources and resolve the discrepancy. The patrol should ensure the area is secure, and the incident should be recorded in the blotter.

P.4 LOST AND FOUND

1. NSF members who recover found property shall tag the items with a DON evidence tag and complete the DON evidence/property custody receipt. The property shall then be relinquished to the lost-and-found custodian.

2. The lost-and-found custodian shall maintain a logbook that contains the recovery date, description of item, and final disposition, including date. Quarterly inventories shall be conducted by a senior enlisted (E-6 or above) or an officer who is not directly involved in the lost-and-found process. Records of inventories shall be recorded in the lost-and-found log book.

3. Lost-and-found property shall be stored in a secure area, separate from evidence.

4. Found property shall be retained for 120 days after attempts to notify the owner have been made. If there is no indication of ownership, the 120 days will commence following the completion of departmental procedures (e.g., posting of notices, advertisements, etc.).

5. Lost-and-found property shall be disposed of IAW with the evidence disposal procedures in Appendix I-8. For those items that could be used by an installation charitable organization (bicycles, tricycles, etc.), the SO must coordinate transfer with the SJA.

P.5 PROPERTY MANAGEMENT

1. Government property may be removed from an installation only by obtaining permission from the CO or designated representative. The supply representative shall obtain the necessary permission from the commander and ensure all other regulatory requirements are met.

2. Securing equipment and government property.

   a. All equipment shall be stored in secure areas when not in use. Regularly used office equipment and machines (computers, printers, adding machines, etc.) shall be monitored by owner-user personnel or shall be locked behind two hard-core doors.

   b. When a supply custodian is expected to be absent for 30 days or more or unexpectedly becomes absent, the alternate supply custodian shall complete an inventory and assume responsibility for securing and accounting for the government property involved.

3. Missing or unaccounted-for government property. A Report of Survey (DD Form 200) shall be initiated within seven days after a local organization-level investigation indicates a loss of any equipment item.

4. Personal property shall be clearly marked on the exterior (engraved or indelible marker) to indicate the owner. The location of the marking is up to the owner and it need not deface the item or be located in a place that would tend to devalue the item. Personal property will be secured in the same manner as government property.

P.6 CONTROLLED AREAS AND CIRCULATION CONTROL

1. Each department shall develop procedures to ensure the proper security and protection of its controlled areas.

2. The owner-user shall provide a list to the SO of the designated controlled area, ECPs, and a point of contact.

3. The SO shall maintain a list (labeled FOR OFFICIAL USE ONLY) of all controlled area points/gates, the owner-user, and a primary/alternate point of contact with duty and emergency contact numbers, as well as all gate combinations.

4. Entry control management for controlled areas.

   a. All gates shall be secured when not in regular use.

   b. Owner-user personnel are responsible to ensure all personnel (pedestrians or in vehicles) using the gates are authorized in the controlled areas.

   c. Gates must be secured after each opening.

P.7 KEY AND COMBINATION LOCK CONTROL PROGRAM

P.7.1 General

Navy activities shall establish a key and lock control program for all keys, locks, padlocks, and locking devices used to meet security and loss-prevention objectives. Security requirements for sensitive locks and keys shall comply with published guidance.

1. Keys securing government property and facilities shall be controlled at all times. A list of all keys and what they open shall be labeled FOR OFFICIAL USE ONLY and secured in a lockable container (key lock box, lockable desk drawer, file cabinet, etc.) separate from the container where the keys are secured.

2. This requirement does not include keys, locks, and padlocks used for convenience, privacy, or administrative or personal use.

3. Combination locks and procedures shall also be controlled. After being unlocked, locks shall be resecured immediately on an opened or closed hasp as required by the situation, and the dials shall be turned to prevent substitution or compromise of the combination. To ensure integrity of combinations, they shall be changed periodically as personnel depart. The CO will determine the frequency and necessity of changing combinations on a case-by-case basis.

P.7.2 Arms, Ammunition, and Explosives

Approved locks, locking devices, and storage procedures that meet security standards in OPNAVINST 5530.13 (series), Department of the Navy Physical Security Instruction for Conventional Arms, Ammunition, and Explosives (AA&E), shall be used.

1. The CO or designee must appoint, in writing, an AA&E key and lock or access control custodian who will ensure proper custody and handling of AA&E keys and locks. The custodian may be assigned responsibility for all locks and keys or just those to AA&E spaces. Key custodians shall not be unit armorers.

2. Approved locks, locking devices, and storage procedures that meet security standards in OPNAVINST 5530.13 (series), Department of the Navy Physical Security Instructions for Conventional Arms, Ammunition, and Explosives (AA&E), shall be used.

   a. Entry doors to armories and magazines must be secured with high-security locking devices (see OPNAVINST 5530.13 (series), Appendix D). Interior doorways may use GSA-approved Class 5 or 8 vault doors. Keyed-alike locks may be used on arms racks.

   b. Doors not normally used for entry must be secured from the inside with locking bars, dead bolts, or padlocks. Panic hardware, when required, must be installed so as to prevent opening the door by drilling a hole and/or fishing from the outside. Panic hardware must meet safety, fire, and building codes and be approved by the Underwriters Laboratory or, when applicable, meet host country requirements.

c. High-security locks or lock cores should be rotated annually to allow preventive maintenance. Secure replacement or reserve locks, cores, and keys to prevent access by unauthorized individuals.

3. Maintain a key control register to ensure continuous accountability of keys. The register must contain the name and signature of the individual receiving the key, date and hour of issuance, serial number or other identifying information of the key, signature of the person issuing the key, date and hour key was returned, and the signature of the individual receiving the returned key. Retain completed registers for three years.

   a. Keep a current roster of these individuals within the unit or activity, protected from public view. Keep the number of keys to a minimum.

   b. Keys must never be left unattended or unsecured. When not attended or in use, that is, in the physical possession of authorized personnel:

      (1) Keys to Category I and II AA&E must be stored in GSA-approved Class 5 security containers or weapons storage containers.

      (2) Keys to Category III and IV AA&E may be secured in containers of at least 12-gauge steel or equivalent (other existing containers may continue to be used). These containers must be secured with a GSA-approved, built-in, three-position, changeable-combination lock; a built-in combination lock meeting UL Standard 768 Group I; or a GSA-approved, key-operated padlock.

   c. If keys are lost, misplaced, or stolen, replace the affected locks or cores immediately. Also secure replacement or reserve locks, cylinders, and keys to prevent unauthorized access to them.

   d. Inventory keys and locks semiannually and keep inventory records for three years.

   e. Locks protecting AA&E spaces may not be part of a master key system.

   f. Maintain keys to AA&E and IDSs separately from other keys and allow access only to those individuals whose official duties require it.

P.7.3 Frequency of Inventories

The frequency of inventories shall be IAW applicable policy or semiannually.

P.7.4 Padlocks and Lock Cores

Where padlocks and removable lock cores are used, there will be a program to rotate these locks and cores annually. The intent is that anyone possessing a key without authorization eventually discovers that the location of the lock that the key fits is no longer known.

P.7.5 Lock Procurement

The SO should be involved in the lock procurement process so that only locks that are adequate for their intended application are procured.

P.7.6 Lockouts

All lockouts involving locks used to meet security objectives shall be promptly examined by competent personnel to determine the cause of the lockout, and the SO should be notified of the determination.

P.8 LOSS PREVENTION SURVEYS AND SECURITY CHECKS

1. In addition to all normally inhabited facilities, surveys shall be conducted at the following facilities.

   a. AA&E storage facilities.

   b. Exchange and commissary facilities.

   c. Storage facilities containing sensitive and/or high-value materials.

   d. Activities possessing restricted areas and facilities and mission-critical areas designated in writing by the commander.

   e. The office, headquarters, or residence (if applicable) of the installation command staff and any local or regional senior leadership that regularly works or resides on the installation.

   f. Other Navy activities not located on a Navy installation may request NSF PS surveys of their facilities. The RSO will coordinate such requests.

2. Security checks.

   a. Each activity shall establish a system for the daily after-hours checks of restricted areas, facilities, containers, and barrier or building ingress and egress points to detect any deficiencies or violations of security standards.

   b. Security deficiencies or violations found during after-hours checks must be reported to the SO within 24 hours of the discovery and must be reported by desk journal entries.

   c. Records of security violations detected by NSF personnel shall be maintained for a period of three years. The SO must follow up each deficiency or violation and keep a record for a period of three years of all actions taken (structural, security, disciplinary, administrative, etc.) by the responsible department to resolve the present deficiency or violation and to prevent recurrence.

P.9 HIGH-RISK AND HIGH-VALUE MATERIAL CONTROL

P.9.1 Security of High-Value Materials

High-value and controlled inventory items (e.g., repair parts for ship and aircraft, tools, controlled substances, computers, and processors) shall be marked and/or have characteristics so that they can be identified, accounted for, secured, or segregated to ensure their protection and integrity. Special attention shall be paid to the safeguarding of inventory items by judiciously implementing and monitoring PS measures. This process shall include analysis of loss rates through inventories, reports of surveys, and criminal incident reports to establish whether repetitive losses indicate criminal or negligent activity.

P.9.2 Command Responsibilities

1. Establish PS measures to protect inventory items and reduce the incentive and opportunity for theft.

2. Monitor the effective implementation of security requirements through scheduled inspections of and staff or oversight visits to affected activities.

3. Ensure security considerations are incorporated into the construction of storage facilities (i.e., vaults or strong rooms) for high-risk and high-value items.

   a. If storage containers are used, they must meet prescribed GSA size and strength requirements.

b. Some containers may require being anchored to the floor or wall to meet security standards.

c. Security containers should also be located within a continuously manned space or be checked by a security force member at least twice per 8-hour shift.

4. Inventory. Ensure accountability and inventory of all high-value and high-risk items are accomplished as prescribed in their particular reference series (e.g., AA&E, maintenance, or medical).

APPENDIX Q

Responding to Bomb Threats

Q.1 OVERVIEW

The use of IEDs can enhance the violence that gives terrorist groups their ability to intimidate or coerce a target population. The detonation itself creates a highly visual, newsworthy scene, even hours after the detonation occurs. Bombs can detonate anywhere, without apparent reason or warning. The use of bombs in a terror campaign emphasizes the authorities' inability to safeguard the public and maintain law and order. Bombs are ideal weapons because they can be designed to give terrorists opportunities to escape from the scene of their crimes.

This procedure identifies information to be obtained from the complainant, safety precautions for handling a bomb threat call, and techniques of searching for a bomb. EOD teams shall examine suspect packages or containers.

The dispatcher shall obtain all pertinent information regarding the person who made the threat. The person receiving the call will fill out the Navy's Bomb Threat Information Form (OPNAV Form 5527/8). The NSF will then use this information to determine whether the affected area warrants a bomb search.

Q.2 RESPONSIBILITY AT THE SCENE

The first NSF to arrive at the scene may leave their handheld radios, mobile phones, and/or pagers on.

WARNING

Once within 50 feet of the building, all communication devices must be secured
and no communications transmissions allowed, because the transmitting
frequency might trigger the explosive device.

The NSF will then contact the building's owner/person in authority to determine whether the building should be evacuated. As soon as possible, the responding NSF shall inform the shift supervisor of the situation. The shift supervisor should contact an EDD team, if available, to assist in the search of buildings, vehicles, and open areas adjacent to buildings and parking lots.

1. If the building is to be evacuated:

   a. The patrol supervisor shall request additional assistance if needed (telephone preferred).

   b. The patrol supervisor shall request the person in authority to solicit building occupants to assist in searching the building (maintenance and building engineers preferred).

2. If the building is not evacuated:

   a. The shift supervisor shall obtain permission from the person in authority to conduct a discreet search of the premises, especially restrooms and areas that are accessible to the public.

   b. The patrol supervisor shall solicit the help of maintenance personnel and building engineers to assist in the search.

    c. The responding NSF should advise all people involved not to answer phones, turn light switches on or off, or do anything that might activate an explosive device.

    d. Should a suspected explosive or incendiary device be located, the NSF shall prevent any unauthorized persons from touching it and ensure that the area is evacuated.

Q.3 SEARCHING FOR THE BOMB

1. Depending on location and manpower allocated, NSF should establish an ICP. The keys to a successful ICP are flexibility and mobility. Once established, the ICP normally does not move except for safety reasons.

2. The best people to perform the search are employees familiar with the area. They are more aware of items that are out of place or alien to the location. The ideal team is two—one employee, one NSF.

3. Actions by those participating in the search should be well planned by the shift supervisor; there should be no undue movement of items, bumping, or shaking. A bomb can be any shape, size, or color.

4. If a suspected bomb is found in an occupied building, the NSF shall evacuate the building.

Q.4 SEARCHING BUILDINGS

The following guidelines should be followed in conducting a bomb search in a building or an automobile:

1. Start outside and work inside.

2. Start at the level most accessible to the public (usually the ground floor).

3. The search should be broken into three steps:

    a. Exterior: The exterior search begins at the ground level. Close attention should be given to piles of leaves and refuse, shrubbery, trash cans, and parked vehicles (outside accessibility is unlimited).

    b. Public area search: Extended outward from the building to some natural divider (curb or wall, usually 25–50 feet).

    c. Interior room search:

        (1) Whenever first entering a room, remain completely calm and immobile. Listen for any unusual sounds. Many times such actions will pick up sounds indicating a device.

        (2) Special attention should be given to utility rooms or areas where access is unlimited.

        (3) Begin the search at the lowest level and work upward, completely searching each level before changing floors.

        (4) When searching a room, search in four levels:

            (a) Floor (and subfloor) to waist: Check chairs, desks, trash cans, anything in this level that could conceal a bomb

            (b) Waist to eye level or top of head: Behind pictures, cabinets, miscellaneous on walls

            (c) From eye level to the ceiling: Light fixtures, any item suspended from the ceiling (e.g., heating and air-conditioning ducts)

            (d) Ceilings and false ceilings.

Q.5 VEHICLE SEARCHES

1. Look for devices designed to kill rather than harass.

2. Whereas a bomb in a building is normally set to detonate at a specific time, an auto bomb usually has a triggering device.

3. The initial action in a vehicle-involved incident is to evaluate and secure as much information as possible from the driver/owner (any threats, evidence of tampering, suspicious noises, unfamiliar objects).

4. Find out when the vehicle was last operated; whether it is locked; and who may be the next person in the vehicle (who is the intended target, driver, or passenger?).

5. Check the area around the vehicle for signs of tampering (marks on the ground, bits of tape, wire insulation, etc.).

Note

A vehicle bomb can be installed in 15–30 seconds. A detailed search emphasizing safety may take several hours depending on the situation.

Q.6 LOCATING SUSPECTED IMPROVISED EXPLOSIVE DEVICES

1. If a suspected bomb is located, it shall remain untouched. The patrol NSF shall notify dispatch and request that the bomb technician be dispatched to the scene. At no time shall any NSF attempt to move or render safe any suspected explosive device.

2. The NSF on the scene shall then evacuate the area.

3. The dispatcher shall also advise other emergency equipment and services be dispatched to the scene.

4. Upon arrival, the bomb technician and/or EOD authority shall supersede all other NSF.

Q.7 CONCLUSION OF SEARCH

The incident commander at the scene shall notify the owner/person in authority of the building of the results of the search. The incident commander shall make sure that all required reports have been completed and all units are back in service.

INTENTIONALLY BLANK

APPENDIX R

Communication Procedures and Plan Requirements

R.1 NAVAL SECURITY FORCES COMMUNICATIONS OVERVIEW

The NSF uses standardized radio communication practices that are compatible with most other LE agencies. The NSF uses both wired and radio equipment to maintain continuous two-way voice communications among all elements of the security force. Two-way radios are the primary means of communications between the dispatcher and field personnel. Permanent-fixed posts are always provided with at least two means to communicate with emergency dispatch or other posts/patrols. Mobile patrols typically have both a multiple-frequency radio and mobile telephone.

1. Routine NSF duties require effective coordination and communication with other police agencies, fire departments, EMS, Navy activities, and public service organizations. NSF must be outfitted with communications equipment that provides effective communications not only between the NSF members, but also with those emergency responders and security activities with whom they may interact with in any given situation.

2. As Navy commands replace aging equipment and adopt new technologies, they must ensure that essential communication links exist within their safety, security, and service activities that permit units from multiple activities to interact with one another and to exchange information according to a prescribed method to achieve predictable results. These methods and results will be contained in communications annexes to AT Plans.

3. To ensure cost-effectiveness, efficiency, and the greatest possible interoperability with federal, state, and local community emergency responders, Navy activities should strongly consider the purchase of communications assets that meet established digital standards for wireless communications users and is compatible with current local agency equipment.

R.2 COMMUNICATIONS PROCEDURES AND OPERATIONS

1. Communications provide essential support for day-to-day and emergency security and LE activities. Communications are maintained to provide mutual aid and interoperability for coordination between multiple agencies under all conditions. NSF maintain shore-to-ship security communications that allow communication with shipboard security posts supporting waterside security.

2. It is critical for NSF to know the full range of communications procedures, including standard communications, response codes, as well as duress codes to alert all security forces of duress situations.

3. Operational tests of all communication circuits will be conducted daily to ensure they are operating properly.

R.3 COMMUNICATION PLAN AND EQUIPMENT

1. The NSF member must be familiar with the communications plan, including:

   a. Post recognition

   b. Caller recognition

c. Emergency telephone number plan, including 911

d. Alarms: siren, voice, and sensor

e. Data transmission equipment—LE terminals

f. Tactical/disaster response/recovery communications

g. Recall procedures

h. Backup communications.

2. Communications Equipment. All NSF members must be familiar with all assigned communications equipment located at their post or within their AO. Equipment requirements for NSF support include:

a. Base/repeater station equipment

b. Mobile radio equipment

c. Portable radio equipment

d. Primary dispatch communications

e. Sirens/voice alarms.

R.4 RESPONSE CODES

The use of plain language during radio communications is required. The standard 10 code should not be used for radio communications. The elimination of standard 10 codes is to support the Incident Command System and the interaction of NSF with other emergency responders that may not be familiar with law enforcement 10 codes. However, when on patrol, NSF must be familiar with response codes, often called brevity codes, that allow succinct and clear patrol communications and minimize air time, which is often crucial in emergency conditions.

1. Code 1 (Routine). When a call is not given a priority code, assume it is routine.

a. Respond by observing all applicable traffic laws.

b. Never use emergency lights or siren for any routine call.

c. If NSF become aware of circumstances unknown to the dispatching agency, they may upgrade the response to Code 2 or Code 3. The responsibility for upgrading the call in this fashion rests solely with the responding patrol.

2. Code 2 (Urgent). A call requiring an immediate response to a nonlife-threatening emergency is normally assigned an "urgent" priority.

a. Respond by observing all applicable traffic laws.

b. Use emergency lights for all urgent calls.

c. Sirens are not authorized.

d. The urgent call is also known as the "silent response." Use this type of response when answering nonlife-threatening, crime-in-progress calls. Check local, state, territorial, or HN traffic codes for limitations on use of lights and siren (some traffic codes do not support Code 2 responses).

3. Code 3 (Emergency). A call requiring an immediate response to a life-threatening emergency or in response to an emergency involving threats to Navy security is normally assigned an "emergency" priority.

   a. The use of emergency lights and siren is mandatory; however, use common sense when approaching the scene of the emergency.

   b. If the emergency lights and siren put security police, victims, or bystanders in peril, turn them off a safe distance from the scene.

R.5 ANTITERRORISM COMMUNICATION REQUIREMENTS

1. COs should ensure that comprehensive communication plan annexes are included in LE, PS, and AT plans or other relevant plans. These annexes will include the concepts of emergency communications, a frequency plan, system requirements, and equipment requirements for the NSF.

   a. Concepts of emergency communications, including:

      (1) Public access and emergency notification.

         (a) Emergency telephone number plan.

         (b) Caller recognition.

         (c) Siren/voice alarms.

         (d) Data transmissions.

         (e) Silent duress alarms.

      (2) Tactical/disaster response/recovery communications.

      (3) Recall procedures.

      (4) Backup communications.

         (a) Use of organic assets.

         (b) Acquisition of external assets.

      (5) Telephone interconnection (with other EOC/command posts and transfer of information via various means (e.g., voice/fax/data).

      (6) Communications reliability.

   b. Frequency plan to include a crypto annex.

   c. System requirements.

      (1) Mobiles and portables.

      (2) C2 communications.

      (3) Mutual aid communications.

    d. Equipment requirements.

        (1) Base/repeater station equipment.

        (2) Mobile radio equipment.

        (3) Portable radio equipment.

        (4) Primary dispatch communications.

        (5) Sirens/voice alarms.

    e. NSF communications.

        (1) The NSF requires sufficient equipment to maintain continuous two-way voice communications among all elements of the security force. Two-way radio will be the primary means of communications between the dispatcher and field personnel.

        (2) Permanent fixed posts will be provided with at least two means to communicate with emergency dispatch or other posts/patrols. Mobile patrols will be provided a multiple frequency radio or radio and mobile telephone.

        (3) Access to communications areas (emergency dispatch center) shall be strictly controlled and protected as a Level 2 restricted area.

        (4) Communications procedures will include provisions for a duress code (changed frequently, immediately if compromised) to alert all security forces of duress situations.

        (5) Interoperability will exist between NSF and other emergency responders. Day-to-day mutual aid and task force interoperability will be addressed.

            (a) Day-to-day interoperability will address those tasks which are commonly encountered and typically associated with areas of coordinated PS/LE responsibilities where Navy activities need to monitor each other's routine traffic. It will also address the need for NSF to monitor local agencies' communications in areas under concurrent jurisdiction or where the Navy possesses only proprietary jurisdiction and primarily relies on local authorities for LE services.

            (b) Mutual aid interoperability will address the requirement for coordination between multiple agencies under conditions that allow little prior planning for the specific event (i.e., natural or man-made disaster).

            (c) Task force operations will detail communications among activities/agencies representing several units and/or layers of government under conditions that do allow for prior planning.

        (6) NSF at installations possessing docking facilities will possess shore-to-ship security communications that allow communication with shipboard security posts watching over waterside security.

2. Operational tests of all communication circuits will be conducted daily to ensure they are operating properly. Electronics personnel will conduct maintenance inspections of all communications equipment periodically.

APPENDIX S

Watch-Standing Procedures

S.1 GENERAL

The RSO of each region should maintain an electronic copy of each installation's submitted watch standing or post orders. From this point forward these orders will be referred to as post orders. These post orders will pertain to each fixed and mobile post in the installation's unit RRP. All post orders will be developed by the SO and approved by the CO prior to forwarding to the RSO for submission.

1. All orders will specify the limits of the post; the hours the post is to be manned; the special orders, duties, uniform, arms (including fields of fire), weapon conditions and equipment prescribed for members of the security force; and detail information for changes of all FPCONs. Additionally, all orders will contain guidance on the use of force as outlined in DODD 5210.56, Carrying of Firearms and the Use of Force by DoD Personnel Engaged in Security, Law and Order, or Counterintelligence Activities. If specified preplanned responses are required for a particular post, they will additionally be included in the post order.

2. The SO will conduct an annual review of all security force post orders at least every 12 months and sign and retain a master copy. The SO will ensure current copies of the signed orders are always maintained for each post.

Figure S-1 is an example fixed/mobile post order, which should be tailored for the individual requirements of the post and installation.

Sample Fixed/Mobile Post Order

Call Sign: XX Gate

Duty Title: Installation Entry Controller, XX Gate

Primary Duties: The ECP is the first point of contact for those seeking access and is the most critical part in the installation's defense-in-depth. Allow entry to personnel who possess valid DOD-affiliated identification credential with an affixed DD Form 2220 vehicle decal. Personnel with a rental vehicle who request installation entry must present an authorized rental agreement with a valid DOD-affiliated ID card or must possess an authorized visitor pass. Unescorted entry will also be granted to civilian law enforcement, medical, or fire department agencies performing official duties only. Personnel who do not possess valid identification will be escorted by authorized personnel possessing a valid DOD-affiliated ID card. Those personnel who wish to gain access to the installation without the above criteria will be denied entry. You will maintain professional posture and bearing while performing entry control duties. You are the first impression people see of the installation.

Post Limits: Not to exceed a distance which you can effectively control vehicular and pedestrian traffic to prevent unauthorized entry.

Hours of Operation: This post is open during the following FPCONs:

FPCONs NORMAL–BRAVO: 24 hours

FPCON CHARLIE: 0600–1700

FPCON DELTA: 0600–1500

FPCON CHANGES: This post is manned IAW the Post Manning Chart.

FPCONs NORMAL–BRAVO: 1 NSF (M-9)

FPCON CHARLIE: 2 NSF and 1 Overwatch position (M-16/M-4)

FPCON DELTA: 2 NSF and 2 Overwatch positions (M-16/M-4)

Armament: M-9 with ball ammunition. Secondary use-of-force capabilities are the authorized baton and oleoresin capsicum (OC) spray chemical dispenser (if qualified) for nonlethal force capabilities.

Required Equipment: Prescribed uniform required IAW Navy instructions. Specified required equipment IAW the CNIC Table of Allowances. Any required restricted area badges or ID cards, handcuffs, two-way portable radio, operational flashlight (all shifts), whistle, appropriate inclement weather gear, and standardized web gear. Ballistic vests will be worn (if issued) at all times while on post. The helmet and gas mask (if issued) will be readily available on post. During FPCON Charlie or higher, the flak vest and helmet will be worn per the direction of down-channeled orders and as prescribed by the installation security officer.

Communication: Two-way portable radio, land lines, flashlight, whistle, and manual (hand and arm) signals.

Figure S-1. Fixed/Mobile Post Order Example (Sheet 1 of 3)

General Instructions: Upon assuming post, you will:

1. Thoroughly inspect your post and report discrepancies to dispatch.

2. Ensure that the gatehouse and surrounding area are kept neat and clean at all times.

3. Assist motorists requesting directions. If motorists allowed entry impede traffic, have them pull off to the side to allow traffic to progress.

4. Upon request or when answering the telephone, give name and rank. Refer all requests for names of other personnel to the WC. Upon request, give out the WC's name and the precinct's telephone number.

5. Conduct a radio check with the desk every 60 minutes during daylight hours, and every 30 minutes during hours of darkness. During FPCON CHARLIE or higher, conduct status checks every 30 minutes during daylight hours and 15 minutes during hours of darkness or reduced visibility.

6. Immediately conduct communication status checks with the dispatch upon assuming post.

7. Ensure all traffic control devices (i.e., bollards/barriers) are working properly and are in place before relieving off-going personnel.

Specific Instructions:

1. Personnel shall render a hand salute to all commissioned officers entering the installation or vehicles with blue decals.

2. Media representatives will not be permitted entry onto the base without a public affairs sponsor or escort.

3. Do not accept packages, parcels, or keys for holding or future delivery to other people.

4. Portable radios are not authorized on post.

5. Personnel shall not touch any vehicle nor allow any part of their person to enter the vehicle while conducting ID checks except when effecting an apprehension or in the line of duty.

6. Eating a small snack and drinking is allowed out of public view as the workload permits. This includes buildings with one-way windows. No food deliveries are allowed. Smoking is allowed off the defined perimeter of the post but still being watchful of the relieving NSF.

Prohibited Items/Actions:

1. Leaving your post without permission.

2. Reading material (other than post or duty-related material).

3. Playing cards, games, or any items not directly associated with your post.

4. Smoking, using smokeless tobacco, drinking, or gum chewing outside the gatehouse when performing official duties.

Figure S-1. Fixed/Mobile Post Order Example (Sheet 2 of 3)

5. Eating meals at the gate while performing official duties without proper relief.

6. Waving traffic from inside the gate shack. Personnel shall stand outside and be visible as required, while conducting their duties.

7. Handling wallets, handbags, or other personal property when conducting ID checks.

8. Accepting gifts, favors, or gratuities from any person while performing official duties.

9. Allowing unauthorized personnel or personnel not on official business in the gatehouse.

10. Portable television.

Name, Rank, USN
Installation Security Officer

Figure S-1. Fixed/Mobile Post Order Example (Sheet 3 of 3)

Attachment 1. Use of Force
(Reference: DODD 5210.56)

1. In cases warranting the use of force, NSF personnel may only use that force reasonably necessary to control or stop unlawful resistance to reach the objective and prevent the unlawful commission of a serious offense. Always use the following principles.

   a. Force must be reasonable in intensity, duration, and magnitude.

   b. There is no requirement to delay force or sequentially increase force to resolve a situation or threat. NSF personnel will attempt to de-escalate applied force if the situation and circumstances permit. NSF personnel will warn persons and give the opportunity to withdraw or cease threatening actions when the situation or circumstances permit.

   c. Warning shots are prohibited. Warning shots are authorized from U.S. Navy and Naval Service vessels and piers in accordance with Chairman of the Joint Chiefs of Staff Instruction 3121.01B, "Standing Rules of Engagement/Standing Rules for the Use of Force for U.S. Forces," 13 June 2005.

   d. Firearms shall not be fired solely to disable a moving vehicle. When deemed as a threat to DoD assets or persons' lives, NSF personnel shall use reasonably necessary force and caution when firearms are directed at a vehicle borne threat.

   e. If excessive force is used in discharging assigned responsibilities, NSF shall be subject to administrative or judicial action IAW Article 92 of the UCMJ or U.S., local, or HN laws.

2. Special considerations for the use of deadly force. Fire shots only with due regard for the safety of innocent bystanders. Do not fire shots if they are likely to endanger innocent bystanders. When possible, give an order to "HALT" before discharging a firearm to prevent death or serious bodily harm to others.

   a. When you are under hostile attack in protection and recovery operations "involving nuclear weapons or lethal chemical agents," the presence of innocent bystanders or hostages "must not deter" you from stopping the attack through all means at your disposal.

   b. When you discharge a firearm, fire it with the intent of rendering the targeted person or persons incapable of continuing the activity or course of behavior that led you to shoot. Rendering incapable may not require death in every circumstance. Warning shots are/are not authorized at (installation).

   c. Follow the specialized rules of engagement for protection of assets IAW NTTP 3-07.2.1, "Antiterrorism."

Figure S-2. Use-of-Force Model (Sheet 1 of 2)

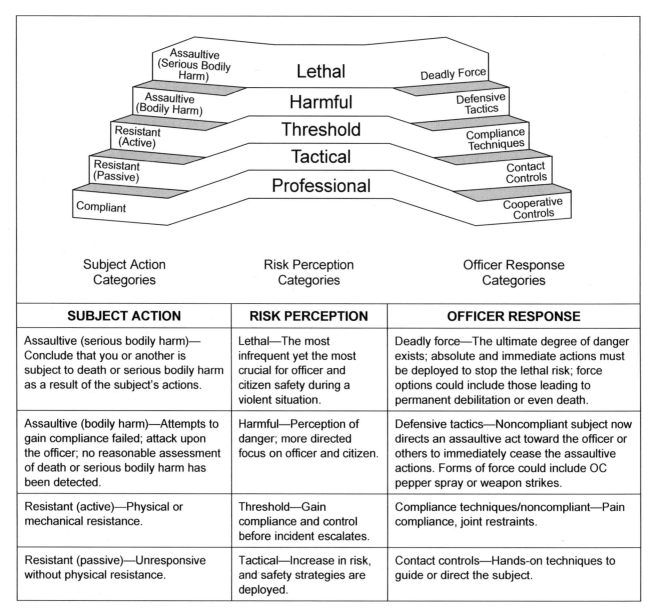

SUBJECT ACTION	RISK PERCEPTION	OFFICER RESPONSE
Assaultive (serious bodily harm)—Conclude that you or another is subject to death or serious bodily harm as a result of the subject's actions.	Lethal—The most infrequent yet the most crucial for officer and citizen safety during a violent situation.	Deadly force—The ultimate degree of danger exists; absolute and immediate actions must be deployed to stop the lethal risk; force options could include those leading to permanent debilitation or even death.
Assaultive (bodily harm)—Attempts to gain compliance failed; attack upon the officer; no reasonable assessment of death or serious bodily harm has been detected.	Harmful—Perception of danger; more directed focus on officer and citizen.	Defensive tactics—Noncompliant subject now directs an assaultive act toward the officer or others to immediately cease the assaultive actions. Forms of force could include OC pepper spray or weapon strikes.
Resistant (active)—Physical or mechanical resistance.	Threshold—Gain compliance and control before incident escalates.	Compliance techniques/noncompliant—Pain compliance, joint restraints.
Resistant (passive)—Unresponsive without physical resistance.	Tactical—Increase in risk, and safety strategies are deployed.	Contact controls—Hands-on techniques to guide or direct the subject.

Figure S-2. Use-of-Force Model (Sheet 2 of 2)

Attachment 2. Post and Weapon Briefings

Procedures: Conduct a post briefing when supervisory or officers within the chain of command approach your post. Also offer post briefings to any Inspector General Staff personnel. Come to the position of attention and report as follows:

Sir/Ma'am, (Rank/Name) reports Gate XX is all in order.

Sir/Ma'am, I am responsible for controlling entry and exit from the installation and give directions to person(s) that require assistance.

My post limits are the distance at which I can effectively control vehicular and pedestrian traffic.

My primary means of communication is the handheld radio. My secondary means of communication consists of land lines, flashlight, whistle, and manual hand and arm signals.

I am armed with the Berretta M-9 pistol and 9 mm ball ammunition. The maximum effective range of my weapon is 50 meters, and the maximum range of my weapon is 1800 meters.

I will use the reasonable amount of force necessary to include deadly force to protect personnel and resources as defined in SECNAVINST 5500.29C.

My 11 General Orders are:

1. To take charge of this post and all government property in view.

2. To walk my post in a military manner, keeping always alert and observing everything that takes place within my sight or hearing.

3. To report all violations of orders I am instructed to enforce.

4. To repeat all calls from posts more distant from the guardhouse (quarterdeck) than my own.

5. To quit my post only when properly relieved.

6. To receive, obey, and pass on to the sentry who relieves me all orders from the commanding officer, command duty officer, officer of the deck, and officers and petty officers of the watch only.

7. To talk to no one except in the line of duty.

8. To give the alarm in case of fire or disorder.

9. To call the officer of the deck in any case not covered by instructions.

10. To salute all officers and all colors and standards not cased.

11. To be especially watchful at night and during the time for challenging, to challenge all persons on or near my post, and to allow no one to pass without proper authority.

We are in FPCON _____.

My area supervisor is (name).

Sir/Ma'am, this concludes my post briefing. Do you have any questions?

Figure S-3. Post and Weapon Briefings Model (Sheet 1 of 2)

M-9 Briefing

The M-9 pistol is a semiautomatic, magazine-fed, recoil-operated, double-action pistol, which fires a 9 mm round. The overall length of the weapon is 8 1/2 inches and its weight with a loaded magazine is 2 1/2 pounds. The maximum effective range is 50 meters. The maximum range is 1800 meters. The weapon has an ambidextrous safety and a firing pin block. The weapon is capable of firing 15 rounds from a loaded magazine before reloading and is currently loaded with 15 rounds of ball ammunition. I also have two additional 15-round magazines in reserve.

Figure S-3. Post and Weapon Briefings Model (Sheet 2 of 2)

APPENDIX T

Handcuff and Subject Transportation

T.1 OVERVIEW

The decision to handcuff is subject to sound professional judgment based on the facts of the specific incident. Security forces most commonly handcuff to affect an apprehension. However, there is no defined criteria that can match judgment. Handcuffing is never automatic unless transporting in a vehicle. Key considerations include the following:

1. Nature of the offense committed

2. Demeanor/violence potential of the suspect (e.g., cooperative, threatening, frightened)

3. Number of suspects involved

4. Controls needed

5. NSF perception of the threat to his/her own personnel safety and that of innocent third parties.

T.2 HANDCUFFING INFORMATION

Handcuffs do restrain free movement but are not a foolproof restraining system. Never think that since a suspect is restrained they are no longer a threat. Use care, common sense, and discretion anytime a suspect is handcuffed.

1. Stepping through cuffs. Many criminals have adapted techniques like stepping through the cuffs to place their hands in front of their body. From this position the suspect could more effectively fight with NSF personnel. To prevent a suspect from stepping through the cuffs, loop the cuffs through the suspect's belt at a point below the small of his/her back prior to securing the second cuff. Remember a suspect with their hands cuffed in front poses a far greater threat. Consider cuffing to the front only if the suspect is pregnant, wounded, or has a physical handicap precluding cuffing behind the back.

2. Fixed objects. Never handcuff a suspect to a fixed object (e.g., a sign post, chain-link fence, vehicle, aircraft seat, etc.). A suspect cuffed to a fixed object may be trapped in the case of an accident or emergency. If a danger to themselves or others, a suspect may be handcuffed to an interrogation desk or bench within the security office.

3. Suspect safety. NSF apprehending a suspect are entirely responsible for their safety. As an example, during transport, place seat belts around suspects to ensure their safety.

T.3 WHEN TO HANDCUFF

The courts consider handcuffing a use of force. The U.S. Supreme Court decision Graham v Connor (1969) held that police officers' decision to use force must be judged from the "perspective of a reasonable officer, within circumstances that are often tense, uncertain, and rapidly evolving." This standard asks whether the NSF actions are objectively reasonable in light of the facts and circumstances confronting them. Analyze all force situations to ensure use of the minimum level of force that will safely protect you from injury. Do not use more force than is necessary to complete an apprehension, yet keep self and others free of danger.

1. Deciding to handcuff. NSF must search all apprehended persons prior to transport. If NSF decide to search a suspect, then they have also decided to handcuff, as handcuffing is an integral part of the search procedures. Remember personal safety and the safety of the suspect are prime considerations. If the circumstances of the apprehension leave any doubt as to personal safety or the safety of the suspect, then handcuff the suspect using the minimum level of force necessary to complete the handcuffing procedure.

2. Assuring control. During an apprehension, NSF may apply handcuffs to ensure control of suspects at the apprehension site. When applying handcuffs, use reasonable levels of force to achieve control of a resisting suspect and maintain control during apprehension and detention.

3. Nonresisting suspects. If NSF decides to use restraining devices on a nonresisting suspect, accomplish the application of handcuffs with reason and without injury. Telling the suspect of NSF intent to apprehend and allowing the suspect to cooperate minimize the risk of injury.

4. Injuries during handcuffing. If a suspect is injured during the handcuffing stage of an apprehension, the first step is to seek immediate medical attention for the suspect. As soon as possible, provide a detailed written statement of the techniques used and action taken by the apprehended suspect that caused the injury.

5. Procedures for applying handcuffs can be found in Chapter 9 of NTTP 3-07.2.1, Antiterrorism.

T.4 TRANSPORTING SUSPECTS

The process of transporting suspects is one of the most dangerous activities in which any officer may engage. It has many times been referred to as the "weakest link in the realm of suspect security." It is generally viewed as the phase of NSF/suspect contact offering the greatest potential for escape. Therefore, the officer must be highly vigilant and adhere to proven procedures and practices. In almost every case of an escape or injury during an escort or transport, the suspect has taken advantage of a procedural lapse on the part of the transporting officer(s). Remember, NSF members must recognize that the ultimate goal of escorting and transporting is to reach the destination free of officer or suspect incident or injury. NSF safety is paramount when transporting suspects in custody. The transporting security forces member(s) will search the suspect for weapons prior to placing them in the transport vehicle and search the vehicle after transport to ensure the suspect has left nothing behind.

T.4.1 Successful Transport

Suspect transport consists of the controlled and secure movement of a suspect(s) in custody from one point to another. Remember transporting is dangerous. Carelessness and the failure to follow established procedures are the major causes of security forces injuries. NSF are both legally and professionally responsible for the safety and security of the subjects. This process should be conducted humanely, professionally, and efficiently with a minimum of public display. Successful transport consists of the following critical components:

1. Preparation of one's attitude toward the serious nature of the activity, a proactive sense of risk assessment, and strict adherence to procedures.

2. Preplan the specific route of the transport.

3. Establish a line of communication to be maintained throughout the transport. Remember to keep the suspect in sight constantly.

4. Increase attention when approaching the destination, since this may be the suspect's last chance to act.

5. Never inform the suspect of more than the required amount of information prior to transport to minimize the risk of preplanning on the part of the suspect.

6. NSF shall personally conduct the search of the suspect to be transported. Assume nothing—verify everything.

7. NSF shall personally apply and check restraint devices prior to transport to ensure security.

8. NSF shall personally search and secure the transport vehicle prior to and after the transport.

T.4.2 Transport Procedures

The transport vehicle integrity is vital. Transport vehicles fall into two categories: standard passenger-type vehicles and those designed and adapted for transporting suspects. Based on frequency of use, the passenger-type vehicle transport procedures should be trained on and demonstrated. However, most of these principles and procedures are generally adaptable to most of the contemporary transport vehicles used.

1. Inspect the transport vehicle prior to transport. The transporting officer's first task is to become totally familiar with the vehicle security devices (shields, screens, radio, etc.) that enhance safety and security. Search the area where the suspect will be seated thoroughly. If the seat can be removed, do so for closer/greater inspection ease and accuracy. Look for locations (seams, tears, crevices, etc.) where a suspect could place contraband or weapons. Remember specialized security devices designed into the vehicle should augment NSF safety.

2. Officer/single-suspect placement in transport vehicles

   a. Once the suspect has been searched and restrained following proper procedures, the suspect should be escorted to the right rear seat. The officer will still secure the suspect while opening the door for the suspect.

   b. During this phase of the escort/transport, the officer must remain to the rear of the suspect in order to prevent a potential attack. The suspect is now advised to first sit on the seat and then bring their legs into the vehicle.

   c. The officer then proceeds to take his left forearm and place it under the suspects chin, verbally directing and slowly pushing his head backward while the officer applies the suspect seat belt.

   d. If available, the supportive transporting officer now enters the vehicle from the left rear and remains seated during transport directly behind the driver. Primary attention for this NSF is directed toward the subject.

3. Officer/two-suspect placement in transport vehicles

   a. Once searched and restrained, the first suspect should be escorted to the right rear passenger door. They should be placed into the vehicle following the previously stated procedures but positioned in the middle part of the rear seat and secured with the seat belt.

   b. The second searched and restrained suspect is then placed into the right rear seat and the seat belt applied.

   c. If available and recommended, the supporting officer now assumes a position in the left rear passenger seat behind the driver.

4. Other considerations. Ideally, the vehicle should have a safety screen between the front and rear seats. When working as a team in a patrol vehicle equipped with a safety screen, place the suspect in the rear seat on the passenger side and the patrol rider in the front passenger seat next to the driver. ALWAYS USE SEAT BELTS for every person in the vehicle.

Note

At no time should a one-person patrol transport more than one suspect if the vehicle is not equipped with a safety screen unless specifically approved by the supervisor on duty.

T.5 OPPOSITE-GENDER TRANSPORTS

When transporting suspects of the opposite gender, ask another NSF, preferably of the same gender as the suspect, to accompany to preclude any charges of impropriety. If a same-gender individual is not available, notify the dispatcher of location, approximate distance to the designated location, odometer reading/starting mileage. The dispatcher shall respond with and log the departure time. Upon arrival at the designated location, notify the dispatcher of arrival and ending mileage. The dispatcher shall respond and log the arrival time.

T.6 HANDCUFF REMOVAL PROCEDURES

Maintain control of the suspect until you determine removal of the handcuffs is appropriate. To remove the ratchet handcuffs from a suspect you must:

1. Place the suspect in the standing, kneeling, or prone position. Approach and use tactical positioning for removal identical to placement.

2. Grab the linking chain or hinge palm up. Once the cuff is released, tell the suspect to rotate the free arm from the cuff slowly and place his/her hand on the back of their head and keep it there until told to do otherwise. Right-handed NSF remove the left cuff first. Left-handed NSF members remove the right cuff first. Regardless of which cuff is removed first, immediately close the ratchet of the removed cuff so the suspect cannot use the open ratchet as a weapon against you.

3. Keeping your weak hand on the linking chain, remove the key and place it in the opposite lock with your strong hand. Repeat the same directions to the suspect for the opposite hand.

Note

Never let go of the linking chain/hinge.

4. With the key in your strong hand, unlock and remove the other cuff, and step back without delay. Immediately close the second cuff so the suspect cannot use the open ratchet as a weapon if they break away.

5. Direct the suspect to remove their hands from the back of their head and release the suspect as instructed by higher or competent authority.

T.7 ESCORT PROCEDURES

Escorting is related most frequently to the movement of suspects on foot. Normally, it is limited in distance, i.e., to and from the transport vehicle, from an area of apprehension to a secure area, etc. It should be remembered that it does normally represent the initial continuation of contact between the officer and suspect, now moving into the phase centered around the transport process. For this reason it should be used as a time to reinforce and/or establish strict adherence to procedures by both participants.

T.7.1 Compliant Subject Escort Position (No Handcuffs)

1. The suspect is approached from the side/rear by the officer(s) who establishes a control hold with one hand on the suspect's elbow and the other hand of the suspect's wrist.

2. The suspect is now guided to the appropriate destination in a relatively safe and secure manner.

T.7.2 Noncompliant Subject Escort Position (No Handcuffs)

1. Once the suspect has been searched thoroughly, the officer(s) moves to the side/rear on the right side of the suspect.

2. The officer rotates the suspect's right hand with the officer's left hand.

3. The officer's right hand is placed on the right elbow of the suspect to stabilize and secure the suspect during escort.

T.7.3 Reverse Escort (Handcuffed)

1. Once the suspect has been thoroughly searched and handcuffed, the officer moves to the left side/rear position of the suspect.

2. The officer now slides his left arm under the suspect's left arm and initiates a reverse wrist-lock position on the suspect's left hand/arm.

3. The officer can additionally reinforce the suspect's captured left hand and arm with his free right hand and arm, or the officer can secure the suspect and keep his right hand and arm free.

4. The additional positive aspect of this technique is the fact that the suspect is continually destabilized by walking backward throughout the entire escort sequence.

INTENTIONALLY BLANK

APPENDIX U

Waivers and Exceptions Process

U.1 GENERAL

Wherever the mandatory security requirements of OPNAVINST 5530.14 (series), Navy Physical Security and Law Enforcement Program; OPNAVINST 5530.13 (series), Department of the Navy Physical Security Instruction for Conventional Arms, Ammunition, and Explosives (AA&E), or this NTTP cannot be met, installations will request waivers or exceptions IAW this appendix. Waivers and exceptions will be evaluated based on merit only and must include compensatory measures.

1. Requests for waivers and exceptions will be submitted in the format outlined in this appendix.

2. Blanket waivers and exceptions are not authorized. Waivers and long-term exceptions are self-canceling on the expiration dates stated in the approval letters unless CNO (N46) approves extensions. Cancellations do not require CNO approval.

3. The approval level of waivers and exceptions as outlined in OPNAVINST 5530.14 (series), Navy Physical Security and Law Enforcement Program, is depicted in figure U-1. The region and budget submitting office (BSO) must be informed and provide comment on all waivers and exceptions. Each organization in the routing chain must provide an approval or disapproval recommendation.

4. All waivers and long-term exception requests require an itemized cost estimate to resolve the security vulnerability.

U.2 WAIVER EXCEPTION PROCESS

Requests for waivers or exceptions will be submitted as follows:

1. Requests for waivers of specific requirements will be submitted via the chain of command from installation to region to BSO (e.g., CNIC, BUMED, etc.) to Echelon 2 commands. Echelon II commands are delegated authority to approve initial waivers for subordinate commands and their own headquarters. No further

POLICY REQUIREMENT	WAIVER ROUTING	EXCEPTION ROUTING
OPNAV 5530.14 (series)	Installation to region to BSO (e.g., CNIC, BUMED, etc.) to Echelon 2 (approval authority)	Installation to region to BSO (e.g., CNIC, BUMED, etc.) to Echelon 2 to CNO (OPNAV N4P is approval authority)
OPNAV 5530.13 (series)	Installation to region to naval ordnance safety and security activity to BSO (e.g., CNIC, BUMED, etc.) to Echelon 2 (approval authority)	Installation to region to naval ordnance safety and security activity to BSO (e.g., CNIC, BUMED, etc.) to Echelon 2 to CNO (OPNAV N4P is approval authority)
Note: Endorsements will be based on the merits of force protection only.		

Figure U-1. Waiver and Exception Approval Authority

delegation is authorized. The request for waiver must include a complete description of the problem and compensatory measures or alternative procedures, as appropriate. Approved waivers will normally be for a period of 12 months. Extension of the waiver (normally for 12 months) must be requested via the chain of command and approved by CNO (N46). Waiver extension requests will refer to previous correspondence approving initial and previous extensions, as appropriate. (See figures U-2 through U-4.)

2. Requests for exceptions to specific requirements due to a permanent or long-term (36 months or longer) inability to meet a specific security requirement must be forwarded via the chain of command from installation to Region to BSO to Echelon 2 command to CNO (N46) for consideration. Each exception request shall include a description of the problem and compensatory measures and procedures to be employed. Exception requests will be reviewed and endorsed by each echelon in the chain. The same applies to any requests for extension of previously approved long-term exceptions. Correspondence that requests extension of previously approved long-term exceptions shall include a reference to the initial CNO approving correspondence.

3. In other countries, the HN may have ultimate responsibility for certain aspects of security, such as perimeter security for Navy activities located there, and Navy authorities may not be able to implement certain requirements set forth in this instruction. In those instances, formal exceptions are not required. However, the parent Echelon 2 command must review the situation and determine what, if any, measures are appropriate to take to compensate for measures not allowed by the HN.

4. The initiating command will assign a waiver or exception number per subparagraphs f. and g. below. All information requested below must be provided in waiver, waiver extension, and exception (permanent and long-term) requests. Requests shall be in letter format, and all elements of subparagraphs g., h., or i. will be specifically addressed. Nonapplicable elements will be noted as N/A.

5. Each waiver or exception request shall include the assignment of waiver or exception numbers to provide a unique identification of any given waiver or exception with respect to the activity involved and the initial year of the request. Any request for extension of a previously approved waiver or exception will use the same number assigned to the original waiver or exception approval. Each waiver or exception will be identified as follows:

   a. The first six digits, beginning with the letter "N" for Navy represent the Unit Identification Code (UIC) of the activity initiating the request.

   b. The next digit is either "W" for waiver or "E" for exception.

   c. The next two digits represent the serial number of the request, beginning annually on 1 January with 01. Waiver and exception numbers will run sequentially together (e.g., W01-08 followed by E02-08, then E03-08, W04-08, etc.). This method allows activities in the reviewing chain of command to exercise their discretion to change an exception request to a waiver request, and vice versa, without having to re-coordinate the number with the requesting activity.

   d. Original numbers assigned long-term exceptions and waivers will be used when requesting exception or waiver extensions.

   e. The last two digits identify the calendar year of the request.

   Example: N01234-W01-08:

N	Navy activity
UIC	01234 (Navy UIC)
W	Waiver ("E" for exception)
01	1st waiver (or exception) request of calendar year
08	2008 (year initial waiver/exception requested)

Line 1	Waiver number.
Line 2	Statement of waiver requirement and references to chapter, section, and paragraph in the pertinent policy manual that cite standards that cannot be met.
Line 3	Specific description of condition(s) that caused the need for the waiver and reason(s) why applicable standards in this manual cannot be met.
Line 4	Description of the physical location of affected facilities or areas. Identify structures individually by building number.
Line 5	Identify interim mandatory compensatory measures in effect or planned.
Line 6	Describe the impact on mission and any problems that will interfere with safety or operating requirements if the waiver is not approved.
Line 7	Identify resources, including estimated cost from which budgeting decisions can be made, to eliminate the waiver.
Line 8	Identify actions initiated or planned (local capability or other) to eliminate the waiver and estimated time to complete.
Line 9	Provide point of contact to include name, rank/grade, Defense Switched Network (DSN), and commercial phone numbers.

Figure U-2. Example Format Prescribed for Requests for Waivers

Line 1	Exception number.
Line 2	Statement of long-term exception requirement and references to chapter, section, and paragraph in this manual that cite standards that cannot be met.
Line 3	Specific description of condition(s) that caused the need for the long-term exception and reason(s) why applicable standards in this manual cannot be met.
Line 4	Description of the physical location of affected facilities or areas. Identify structures individually by building number.
Line 5	Identify interim mandatory compensatory measures in effect or planned.
Line 6	Describe the impact on mission and any problems that will interfere with safety or operating requirements if the long-term exception is not approved.
Line 7	Identify resources, including estimated itemized cost from which budgeting decisions can be made, to eliminate the long-term exception.
Line 8	Identify actions initiated or planned (local capability or other) to eliminate the long-term exception and estimated time to complete.
Line 9	Provide point of contact to include name, rank/grade, DSN, and commercial phone numbers.

Figure U-3. Example Format Prescribed for Requests for Long-term Exceptions

Line 1	Exception number.
Line 2	Statement of the exception requirement and reference to the chapter, section, and paragraph in this instruction that cite the standard that cannot be met.
Line 3	Specific description of condition(s) that caused the need for the permanent exception and reason(s) why applicable standards in this manual cannot be met.
Line 4	Description of physical location of affected facilities or areas. Identify structures individually by building number.
Line 5	Identify, in detail, compensatory security measures that are being applied.
Line 6	Describe the impact on mission and any problems that will interfere with safety or operating requirements if the exception is not approved.
Line 7	Provide point of contact to include name, rank/grade, DSN, and commercial phone numbers.

Figure U-4. Example Format Prescribed for Requests for Permanent Exceptions

APPENDIX V

Fingerprint Processes and Procedures

V.1 GENERAL

Criminal fingerprint cards (FD-249) will be submitted for all military service members when a command initiates military judicial proceedings or when command action is taken in a nonjudicial punishment proceeding for the commission of any offense IAW DODI 5505.11, Fingerprint Card and Final Disposition Report Submission Requirements.

1. These procedures must be followed to ensure maximum uniformity and standardization in the timely submission of FD-249 and Final Disposition Reports. This guidance does not apply to the DON fingerprint procedures expressed in SECNAVINST 1640.9 (series), Department of the Navy Corrections Manual, or if an apprehension is based solely on the charge of UCMJ Article 85, Desertion. Submit suspect fingerprint cards on the sole charge of military desertion under the Originating Agency Identifier for the respective deserter information point.

2. The investigating agency that submitted the suspect's fingerprint cards and prepared the investigation report is responsible for the final disposition reporting. LE records include arrest and disposition reports. The FBI recognizes NCIS as a central channeling agency for all criminal history reporting including the FBI forms FD-249 and R-84.

3. The FD-249 and R-84, when annotating apprehension—charges and final disposition (adjudication of apprehension—charges), result in the creation of a computerized criminal history record within the FBI Criminal Justice Information Services (CJIS) Division/NCIS Interstate Identification Index.

4. A record remains in effect until the NCIS headquarters, as the channeling agency serving as the contributor, requests the FBI CJIS Division to expunge the arrest or apprehension record.

V.2 PROCEDURES

1. Obtain fingerprints and all additional information required by the FD-249, Suspect Fingerprint Card, from military suspects under investigation by the NCIS or any other Navy and Marine Corps CID, LE, and police organization for the offenses listed in DODI 5505.11, Fingerprint Card and Final Disposition Report Submission Requirements; or from civilian suspects under investigation by the Navy and Marine Corps CID, LE, and police organization for offenses under their jurisdiction and where they are the lead agency in the investigation.

2. Initiate offender criminal history data records required under OPNAVINST 5530.14 (series), Navy Physical Security and Law Enforcement Program, by preparing two original FD-249s. FBI/Department of Justice (DOJ) Instruction, Guidelines for Preparation of CJIS Division Fingerprint Cards, contains guidelines for FD-249 preparation.

3. All criminal fingerprint cards are to be mailed to NCIS Headquarters, Code 24B3 (FINGERPRINTS), 716 Sicard Street SE, Suite 2000, Washington Navy Yard, DC 20388-5380. Questions may be referred to NCIS Headquarters, Code 24B3, telephone number (202) 685-0013.

4. Submit the FD-249 when a command initiates military judicial proceedings or when command action is taken in nonjudicial proceedings against a military subject investigated by an LE organization for the commission of an offense listed in DODI 5505.11, Fingerprint Card and Final Disposition Report Submission Requirements.

5. Submit FD-249 within 15 days of command initiation of military judicial proceedings or when command action is taken in nonjudicial proceedings against a military subject.

6. Hold the FD-249 and record the final disposition on the FD-249 when the final disposition of the proceedings is anticipated within 60 days of command initiation of military judicial proceedings or of command action in nonjudicial proceedings. Submit to the FBI the R-84 if the final disposition is not recorded on the FD-249.

7. Do not delay submission of the FD-249 where the final disposition of the military judicial or nonjudicial proceedings is not anticipated within 60 days.

8. Record as final disposition either on the FD-249 or R-84, as appropriate, the approval of a request for discharge, retirement, or resignation in lieu of court-martial and/or a finding of lack of mental competence to stand trial.

9. Enter the entire assigned Case Number on the FD-249 under Local Identification/Reference block and on the R-84 under Arrest No. (OCA) block.

10. Describe in commonly understood terms (e.g., murder, rape, robbery, assault, possession of a controlled substance, etc.) or, if not specified in DODI 5505.11, Fingerprint Card and Final Disposition Report Submission Requirements, by commonly understood title when submitting the FD-249 and R-84 arrest-charges. Do not describe offenses solely by reference to a UCMJ punitive article or to the USC or other statutory provision. Investigators must ensure that charges annotated on the FD-249 reflect the actual charges being pursued through court-martial or nonjudicial punishment.

11. Describe in common language (e.g., conviction [include offense(s)], dishonorable discharge, reduction in rank, forfeiture of pay, charges dismissed, etc.) the disposition on the FD-249 or the R-84. The disposition of conviction shall only be reported for crimes prosecuted by trials by general, special, or summary courts-martial yielding a plea or finding of guilty. Record adverse findings stemming from nonjudicial proceedings as nonjudicial disciplinary action followed by the punishment awarded.

V.3 FINAL DISPOSITION REPORTING

1. Report within 15 days after final disposition of judicial or nonjudicial proceedings or the approval of a request for discharge, retirement, or resignation in lieu of court-martial disposition information on the R-84 if it has not been reported on the FD-249. Do not hold the FD-249 pending appellate actions; however, report appellate action affecting the character of an initial disposition if it occurs. The SO will also report dispositions that are exculpatory in nature (e.g., dismissal of charges, acquittal).

2. If the SO refers the case to another DON LE agency or LE agency outside DON, that agency will be responsible for the submission of fingerprint cards and reporting of final disposition. The Installation Security Department will furnish any fingerprint records obtained to the receiving agency.

1

APPENDIX W

Restricted Area Security Measures and Signing and Posting of Installation Boundaries

W.1 SIGNING AND POSTING OF INSTALLATION BOUNDARIES

Commanders shall ensure that signs are posted as prescribed below.

1. Size, placement, and use of any language in addition to English should be appropriate for the stated purpose.

2. All regularly used points of entry at Navy installations, separate activities, and restricted areas will be posted at regularly used points of entry with signs that read as follows:

WARNING

**RESTRICTED AREA—KEEP OUT
AUTHORIZED PERSONNEL ONLY**

**AUTHORIZED ENTRY INTO THIS RESTRICTED AREA CONSTITUTES CONSENT
TO SEARCH OF PERSONNEL AND THE PROPERTY UNDER THEIR CONTROL
INTERNAL SECURITY ACT OF 1950 SECTION 21, 50 USC 797**

---

3. Perimeter boundaries of restricted areas that are composed of barriers such as fences or walls not closed off by a roof or ceiling will be posted at intervals with signs that read as follows:

WARNING

**RESTRICTED AREA
KEEP OUT
Authorized Personnel Only**

---

   a. These signs do not have to be posted along the boundaries of the restricted area where walls form an enclosed box with true floor and true ceiling.

   b. Restricted area signs will not indicate whether the area is a Level One, Two, or Three restricted area.

   c. The issue of whether to post perimeter boundaries of Navy installations and separate activities will be governed by trespass laws applicable to the jurisdiction in which the installation/activity is located. Wherever possible, perimeter boundaries will be posted with signs that read:

**U.S. GOVERNMENT PROPERTY
NO TRESPASSING**

---

   d. Where a language other than English is prevalent, restricted and nonrestricted area warning notices will be posted in both languages.

   e. The interval between signs posted along restricted areas should not exceed 100 feet.

   f. The interval between signs posted along perimeter boundaries should not exceed 200 feet.

W.2 WATER BOUNDARIES

Water boundaries present special security problems. Such areas should be protected by barriers and marked with appropriate signage. In addition to barriers, harbor security boats should be used at activities whose waterfronts contain critical assets. In inclement weather, such patrols cannot provide an adequate degree of protection and should be supplemented by increased waterfront patrols, watchtowers, MWD teams, and other appropriate waterside security systems.

W.3 RESTRICTED AREA SECURITY MEASURES

Commanders shall ensure that the following minimum security measures are employed for restricted areas:

1. Level One restricted areas. This least-secure type of restricted area will be established to provide an increased level of security over that afforded elsewhere aboard the activity, to protect a security interest that, if lost, stolen, compromised, or sabotaged, would cause damage to the command mission or impact on the tactical capability of the United States. It may also serve as a buffer zone for Level Two and Level Three restricted areas, thus providing administrative control, safety, and protection against sabotage, disruption, or potentially threatening acts. Uncontrolled movement within a Level One restricted area may or may not permit access to a security interest or asset. A more in-depth definition of a Level One restricted area can be found in OPNAVINST 5530.14 (series), Navy Physical Security and Law Enforcement Program. For all AA&E facilities, refer to OPNAVINST 5530.13 (series), DON Physical Security Instruction for Conventional Arms, Ammunition, and Explosives (AA&E). A Level One restricted area should have:

   a. A clearly defined protected perimeter.

     (1) Level One restricted areas shall be protected IAW the policy requirements for the security of the types of assets located therein. If no other policy pertains, perimeter protection may consist of a fence or barrier, the exterior walls of a building or structure, or the outside walls of a space within a building or structure. If the perimeter is a fence or barrier, it should be posted at no less than 100-foot intervals along the perimeter. If the perimeter is a wall, it should be posted at the point of ingress.

     (2) If natural barriers form a portion of the Level One restricted area perimeter, they should be augmented with sufficient physical barriers to channel vehicular access to the ECP. Supporting patrols or ESSs should be sufficient to detect any unauthorized entry.

   b. Controlled access accomplished through posted access control personnel or AECS. Cipher locks are not appropriate restricted area access control devices.

     (1) Access control personnel need not be armed and may fulfill this function in conjunction with other duties (i.e., a receptionist). If access control personnel are unarmed or if AECS is used in lieu of an armed sentry, an armed sentry or patrol shall be capable of responding to the area within five minutes of notification of an incident.

     (2) An AECS and CCTV may be used in a manner that does not necessitate a guard's physical presence and allows a single individual to authorize access at multiple ECP.

c. A personnel identification system will be used.

d. Access shall be limited to persons whose duties require access and who have been granted appropriate authorization. Persons not cleared for access to the security interest contained within a restricted area may, with appropriate approval, be admitted, but they must be escorted so that the security interest itself is still protected from unauthorized access.

e. Access list and visit log documentation is required for all noncleared personnel visiting the restricted area. Access documentation for all cleared personnel is required during nonnormal work hours and not required during normal work hours. Such documentation, recording entry and departure and sufficient personal information to be able to identify the visitors in the event of an investigation, may be accomplished through automatic record-keeping logs associated with an AECS. If automated log-keeping systems are used, the AECS shall be programmed to allow access only at prescribed times (e.g., a worker who is not authorized access after normal work hours and attempts access will cause an alarm situation and the attempt automatically recorded). If a computer access control or logging system is used, it should be safeguarded against tampering.

f. Area will be secured during nonwork hours. Cipher locks are not appropriate as restricted area locking devices.

g. Security patrols shall check for signs of attempted or successful unauthorized entry and for other activity which could degrade the security of the restricted area. At a minimum, checks should be made every eight hours during normal work hours and every four hours after normal work hours.

h. A designated response force (consisting of a minimum of two NSF personnel) shall be capable of responding within 15 minutes of notification of an incident.

2. Level Two restricted areas. A Level Two restricted area may be inside a Level One area but will not be inside a Level Three area. It will be established to provide the degree of security necessary to protect against uncontrolled entry into, or unescorted movement within, an area that could permit access to a security interest that, if lost, stolen, compromised, or sabotaged, would cause serious damage to the command mission or gravely harm the operational capability of the United States. Uncontrolled or unescorted movement could permit access to the security interest. At a minimum, Level Two restricted areas will be established around alert systems; forces and facilities; pier facilities for aircraft carriers, submarines, and large deck amphibious ships (priority B); aircraft hangers, ramps, parking aprons, flight lines, runways and aircraft rework areas; all category arms and category I and II ammunition and explosives storage facilities and processing areas (including ammunition supply points); essential command and control, communications, and computer facilities, systems, and antenna sites; critical assets power stations, transformers, master valve, and switch spaces; and research, development, test, and evaluation (RDT&E) centers and assets, the loss, theft, destruction, or misuse of which could impact the operational or tactical capability of the United States. A more in-depth definition of a Level Two restricted area can be found in OPNAVINST 5530.14 (series), Navy Physical Security and Law Enforcement Program. For all AA&E facilities, refer to OPNAVINST 5530.13 (series), DON Physical Security Instruction for Conventional Arms, Ammunition, and Explosives (AA&E). A Level Two restricted area should have:

a. A clearly defined protected perimeter.

(1) Level Two restricted areas shall be protected IAW the policy requirements for the security of the types of assets located therein. If no other policy pertains, perimeter protection may consist of a fence or barrier, the exterior walls of a building or structure, or the outside walls of a space within a building or structure. If the perimeter is a fence or barrier, it should be posted at no less than 100-foot intervals along the perimeter. If the perimeter is a wall, it should be posted at the point of ingress.

(2) If natural barriers form a portion of the Level Two restricted area perimeter, they shall be augmented with sufficient physical barriers to channel vehicular access to the ECP. Supporting patrols or ESSs shall be sufficient to detect any unauthorized entry.

b. Controlled access shall be accomplished through posted armed sentries or AECS. Cipher locks are not appropriate restricted area access control devices.

(1) If AECS is used, an armed sentry or patrol shall be capable of responding to the area within five minutes of notification of an incident.

(2) An AECS and CCTV may be used in a manner that does not necessitate a guard's physical presence and allows a single individual to authorize access at multiple ECPs.

c. A personnel identification system will be used.

d. Access shall be limited to persons whose duties require access and who have been granted appropriate authorization. Persons not cleared for access to the security interest contained within a restricted area may, with appropriate approval, be admitted, but they must be escorted so that the security interest itself is still protected from unauthorized access.

e. Documentation is required for all personnel visiting the restricted area via an access list and visit log. Such documentation, recording entry and departure and sufficient personal information to be able to identify the visitors in the event of an investigation, may be accomplished through automatic record-keeping logs associated with an AECS. If automated log-keeping systems are used, the AECS will be programmed to allow access only at prescribed times (e.g., a worker who is not authorized access after normal work hours and attempts access will cause an alarm situation and the attempt automatically recorded). If a computer access control or logging system is used, it should be safeguarded against tampering.

f. Area will be secured during nonwork hours. Cipher locks are not appropriate restricted area locking devices.

g. Security patrols shall check for signs of attempted or successful unauthorized entry and for other activity that could degrade the security of the restricted area. At a minimum, checks shall be made every eight hours during normal work hours and every four hours after normal work hours.

h. A designated response force (consisting of a minimum of four NSF personnel) shall be capable of responding within 15 minutes of notification of an incident.

3. Level Three restricted areas. This most secure type of restricted area may be within less secure types of restricted areas and will be established to provide a degree of security where access into the restricted area constitutes, or is considered to constitute, actual access to a security interest that, if lost, stolen, compromised or sabotaged, would cause great harm to the command mission or strategic capability of the United States. Access to the Level Three restricted area will constitute actual access to the security interest or asset. At a minimum, Level Three restricted areas will be established around chemical, biological, radiological, and nuclear; and special/nuclear weapons research, testing, storage, and maintenance facilities; critical command and control, communications, and computer facilities, systems and antenna sites; critical intelligence-gathering facilities and systems; nuclear reactors and category I and II special nuclear materials; permanent or temporary pier facilities for fleet ballistic missile submarines armed with nuclear weapons; and RDT&E centers and assets, the loss, theft, destruction or misuse of which would result in great harm to the strategic capability of the United States. A more in-depth definition of a Level Three restricted area can be found in OPNAVINST 5530.14 (series), Navy Physical Security and Law Enforcement Program. For all AA&E facilities, refer to OPNAVINST 5530.13 (series), DON Physical

Security Instruction for Conventional Arms, Ammunition, and Explosives (AA&E). A Level Three restricted area should have:

a. A clearly defined protected perimeter.

    (1) Level Three restricted areas will be protected IAW the policy requirements for the security of the types of assets located therein. If no other policy pertains, the protected perimeter may consist of a fence or barrier, the exterior walls of a building or structure, or the outside walls of a space within a building or structure. If the perimeter is a fence or barrier, it should be posted at no less than 100-foot intervals along the perimeter. If the perimeter is a wall, it should be posted at the point of ingress.

    (2) If natural barriers form a portion of the Level Three restricted area perimeter, they shall be augmented with sufficient physical barriers, ESSs, or continuous security patrols to ensure that all personnel seeking entry, by whatever means, are either denied access or channeled to the designated ECP.

b. Controlled access accomplished through posted armed sentries. Cipher locks are not appropriate Level Three restricted area access control devices.

    (1) An AECS may be used if the area is enclosed within another security enclave to which access is controlled by an armed sentry, present inside the Level Three restricted area to protect actual access to the security asset or interest.

    (2) Under conditions where an AECS may be used, electronic access control systems and CCTV may be used in a manner that does not necessitate a guard's physical presence and allows a single individual to authorize access at multiple ECP.

c. A personnel identification system shall be used.

d. Access shall be restricted to personnel who have duty requirements within and have been authorized in writing by the commanding officer. Persons who have not been cleared for access to the security interest contained within a Level Three restricted area may be admitted to the facility with approval, in writing, from the commanding officer. Such persons and all visitors shall be escorted by an authorized/cleared activity escort at all times and the security interest shall be protected from compromise.

e. Access list and visit log documentation is required for all personnel visiting the restricted area. Such documentation will record entry and departure and sufficient personal information to allow visitor identification in the event of an investigation and may be accomplished through automatic record-keeping logs associated with an AECS. If automated log-keeping systems are used, the AECS will be programmed to allow access only at prescribed times (e.g., a worker who is not authorized access after normal work hours and attempts access will cause an alarm situation and the attempt automatically recorded). If a computer access control or logging system is used, it will be safeguarded against tampering.

f. The area will be secured during nonwork hours. When secured, an electronic security system or security personnel must control access to the area.

g. Dedicated security forces shall continuously check for signs of attempted or successful unauthorized entry and for other activity that could degrade the security of the restricted area.

h. Dedicated security forces (patrols and immediate response forces) should be capable of responding within time limits prescribed in policy and defeating the capabilities of potential adversaries identified in program, system, command and installation threat statements that have assessed the potential threats to the protected security interest.

i. Each Navy activity will establish a system for occupants/users of restricted areas, facilities, containers, and barrier or building entry and departure points to detect any deficiencies or violations of security standards.

INTENTIONALLY BLANK

APPENDIX X

Juvenile Procedures

X.1 GENERAL

The age limits for classifying persons as juveniles vary according to the laws of the particular state. Federal law defines a juvenile as "any person who has not attained his 18th birthday." Active duty military personnel under the age of 18 are subject to the UCMJ and are not affected by juvenile laws. The SO shall become familiar with host-state juvenile statutes and prepare RRP defining actions to be taken with juvenile offenders aboard installation property.

X.2 JUVENILES PROCEDURES

1. Whenever the NSF takes a juvenile into custody for an offense or other act of delinquency, the juvenile will be provided with the appropriate constitutional warning against self-incrimination. This warning must be provided in language that the juvenile can understand.

2. NSF shall notify the parent(s), guardian, or custodian that the juvenile is in custody, the exact nature of the alleged offense, and the juvenile's rights against self-incrimination. This notification must be made immediately after the juvenile is taken into custody and the identification of the responsible adult is obtained. The time of the custody, the time of the notification, and the identity and relationship of the person notified must be reported in the IR. This notification is the responsibility of the detaining agency and must be made even if the matter is referred to the NCIS or other agency. As DON policy, the parent, guardian, or custodian must be given a reasonable opportunity to be present during any interrogation of the juvenile. If the responsible adult cannot be present for any reason, the reason for the absence will be indicated in the IR.

3. Fingerprints or photographs of a juvenile may be taken. These records must be safeguarded under 18 U.S.C. 5038, et. seq. Some states prohibit fingerprinting or photographing of juvenile offenders without a written order of a judge or magistrate. The security officer must consult with the SJA and establish local policy.

4. Federal law and DON policy require that a juvenile in custody be taken before a local magistrate at the first opportunity and that a juvenile not be detained for longer than a reasonable period of time before being brought before a magistrate. Accordingly, juveniles detained by NSF personnel for minor offenses are normally released to the custody of their parents at the earliest opportunity.

5. In many instances, a minor offense can be disposed of without delay. Serious offenses or offenses involving repeat offenders may require administrative or judicial action. The cases require consultation with the SJA.

   a. Offenses under the jurisdiction of NCIS will be referred to NCIS at the earliest possible time. Juveniles detained on offenses under NCIS jurisdiction will be turned over to them at their request, provided there is no unreasonable delay.

   b. Within U.S. jurisdiction, juvenile offenders will be referred to local police juvenile authorities. Security officers will establish liaison with local police to develop working agreements for the referral of juveniles.

6. There are no special requirements concerning the interview of a juvenile as a witness. During the on-scene phase of any incident, juveniles may be interviewed in the same manner as any other witness. As a matter of

policy, however, in-depth or follow-up interviews of juveniles will not be conducted without first advising a parent, guardian, or custodian of the nature of the situation and the need to interview the juvenile.

X.3 RECORDS

1. The age of an offender has no effect on the need for detailed and accurate records of any incident or complaint. An IR will be prepared on each situation that fits the criteria for that form.

2. Security units will establish a separate file for the retention of records concerning juvenile offenders. This file will be in a distinctly different location from adult files to lessen the chance that a juvenile record will be placed in the adult IR files and should be kept locked to prevent unauthorized disclosure. Access to this file will be restricted to individuals specified by the security officer as having a need-to-know. Provided local state law does not object, juvenile records may be released to:

   a. The judge of a juvenile court having jurisdiction over the offender.

   b. An attorney representing the juvenile.

   c. A government attorney involved in the adjudication of the matter.

   d. Another court of law in response to a specific inquiry.

   e. Another LE agency if the request is related to an investigation of a crime or a position within that agency.

   f. The director of a juvenile treatment facility or agency to which the court has committed the juvenile, providing that the request is made in writing.

   g. An agency that is considering the juvenile for a position immediately and directly affecting national security.

   h. The parents or legal guardian of the juvenile offender.

3. Neither the name nor a photograph of any juvenile may be made public.

4. Juvenile courts in the various states may impose additional requirements or restrictions on security units concerning juvenile records. These requirements may include the sealing of records. Any such requests will be coordinated with the SJA.

APPENDIX Y

Department of Navy Consolidated Law Enforcement Operations Center (CLEOC)

Y.1 INTRODUCTION AND OVERVIEW

1. Incident Reporting Requirements.

   a. Criminal complaints, traffic accidents, activities, and significant incidents will be reported via an incident report (IR) using the CLEOC. This web-based program represents the Navy's collection platform for National Incident Based Reporting System (NIBRS) and Defense Incident-Based Reporting System (DIBRS) crime and incident reporting.

   b. Criminal complaints and incidents investigated by a security unit's Command Criminal Investigator (CCI) or Criminal Investigations Division (CID) will be reported using the CLEOC Report of Investigation (ROI).

   c. Security Officers will direct the preparation of an IR or an ROI for every criminal complaint or significant incident that is brought to the attention of the security unit. Complaints against members of the security unit, on or off duty, will also be documented on an IR or ROI as appropriate.

2. The Navy Forms Guide may be used as a reference for the completion of PS and LE forms. Matters in the following categories will be reported on the documents indicated, and will not be reported by IR:

   a. Nonsignificant incidents recorded in the DON Desk Journal will be recorded using CLEOC's Desk Journal module, e.g., such as funds escorts, traffic control operations, etc.

   b. The Vehicle Registration module in CLEOC will be used to document the security unit's vehicle registration activities.

   c. Contact with individuals under suspicious circumstances, but with no immediate indication of criminal activity recorded on the Field Interview Card (OPNAV 5580/21).

   d. Impounding abandoned motor vehicles, or voluntary storage of motor vehicles on the Vehicle Report (OPNAV 5580/12).

Y.2 CLEOC

NSF units will use the CLEOC web-based reporting system to support higher headquarters and to comply with mandated NIBRS and DIBRS reporting requirements. CLEOC is a web-based program, accessible via the internet at https://www.usncleoc.ncis.navy.mil. CLEOC user guides are available on the CLEOC homepage.

Y.3 DISSEMINATION

1. The original IR or ROI and exhibit(s), if any, that report on investigations under jurisdiction of NCIS shall be forwarded to the NCIS office to which the incident was referred and assumed.

2. All requests for copies of IRs or ROIs, other than those by internal command officials, should be directed to region/installation SJA or to the Regional Freedom of Information Act Release Authority. Security Officers must comply with SECNAVINST 5211.5E, Department of the Navy Privacy Act Program.

Y.4 INCIDENT REPORT RETENTION

1. Copies of IRs will be retained by the initiating security Unit for a period of two years from the date of the incident or until all legal appeals have been exhausted. The regional security officer or the local security officer on a case-by-case basis may authorize retention of IRs beyond two years. Any retention of an IR or ROI beyond the two-year limitation must contain a written explanation for the retention, signed by the security officer. IRs stored or archived electronically may be retained for five years if the security Unit has sufficient capacity.

2. The NCIS serves as the centralized repository for law enforcement reports within the DON and as such they maintain reports for periods up to 50 years. Security Units will forward original IRs or ROIs reporting violations referenced in Figure Y-1, Table of Offenses. These IRs and exhibits of incidents that have been adjudicated, or otherwise investigated to the fullest with no apprehension or identification of subjects will be mailed to NCIS Headquarters. These IRs will be mailed even if they are capable of electronic submission of reports. A report will not be forwarded until the incident is closed. Reports forwarded to NCIS Headquarters must be complete and contain all exhibits.

3. Records, criminal reports of investigation, complaint/incident reports, alphabetical index cards and other sensitive information will be safeguarded against unauthorized alteration, destruction or disclosure. Disposition guidance for IRs is contained in SECNAVINST 5210.8D, Department of the Navy Records Management Program.

4. Copies of IRs or ROIs to support official government business may be obtained by mailing a letter request to Naval Criminal Investigative Service Headquarters (Code 11C), 716 Sicard Street SE, Suite 2000, Washington Navy Yard DC 20388-5380. The request must be on the unit's letterhead; contain the subject's full name and personal identifying data (i.e., SSN, date, and place of birth), to the extent known; list the name of a point of contact and telephone number of the requesting official; the purpose of the request (e.g., law enforcement investigation) and be signed by a person in a position of authority. Letter requests may be sent by facsimile to (202) 433-9518.

5. Copies of NCIS provided IRs or ROIs may not be further disseminated without the expressed permission of NCIS.

Y.5 CASE CATEGORIES

Case Categories to be forwarded to NCIS Headquarters (CODE 23C).

Original IRs and exhibits that report violations that fall within the case categories and punitive Articles of the UCMJ and identified in Figure Y-1, Table of Offenses will be forwarded to NCIS Headquarters. No later than the 15th of each month, installation and ship security departments will forward original completed (closed) IR(s) or ROI(s), to include exhibits and case disposition, from the previous month and which pertain to the offenses listed in the Table of Offenses. All existing reports that have not been submitted and meet this criterion must also be forwarded. Reports must be legible to ensure clarity during optical imaging. Mail completed IRs or ROIs to:

IR PROJECT
NAVAL CRIMINAL INVESTIGATIVE SERVICE HEADQUARTERS (CODE 11C)
716 SICARD STREET SE, SUITE 2000
WASHINGTON NAVY YARD DC 20388-5380

Y.6 HATE/BIAS CRIME REPORTING

Patrol officers conducting an investigation that indicates that the offense was committed as a result of hate or bias must annotate it in the appropriate area of the IR or ROI. All hate crimes will be referred to and assumed by NCIS.

Y.7 VICTIM AND WITNESS ASSISTANCE PROGRAM

The Navy's Victim/Witness Assistance Program requires patrol officers and investigators to advise crime victims and witnesses of their rights. These rights are enumerated in SECNAVINST 5800.11 (series), Victim and Witness Assistance Program, and OPNAVINST 5800.7 (series), Victim and Witness Assistance Program.

1. An Initial Information for Victims and Witnesses of Crime (DD 2701) will be issued and annotated with the date issued under the Witness/Victim Information section of the CLEOC Involved Persons screen.

2. Victims/witnesses referred to CCIs or NCIS will have the current telephone number, recommended point of contact and other relevant information completed, in the spaces provided by CLEOC.

3. Patrol officers and investigators should follow local instructions regarding who is eligible for victim/witness compensation. Not all victims and witnesses will be eligible since state boards set limits on the type of crime, extent of loss, and other rules regarding compensation.

Arson	Homicide, Negligent
Assault	Manslaughter
Assault, Intimidation	Homicide, Justifiable
Assault, Simple Assault	Housebreaking
Assault, Aggravated Assault	Hit and Run
Assault, Simple Assault (Constructive)	Indecent acts or liberties with a child
Attempted Suicide	Indecent Exposure
Breaking and Entering	Indecent Language (communicating to any child under the age of 16 years)
Bribery	Indecent acts with another
Burglary	Kidnapping/Abduction
Bad Check ($100 or more only)	Liquor Law Violations
Burning with the intent to defraud	Loitering
Bomb Threat/Hoax	Larceny
Counterfeiting	Mutiny or sedition
Curfew	Murder
Child Abuse	Manslaughter
Controlled Substance Violation	Maiming
Computer related and associated crime	Mails (Taking, opening, secreting, destroying or stealing)
Destruction of Property	Misprison of serious offense
Disorderly Conduct	Property Damage
Driving Under the Influence	Prostitution
Drug equipment/paraphernalia	Prostitution, Assisting or Promoting
Drug/Narcotic Offenses	Perjury
Drunkenness	Public Record (Altering, concealing, removing or mutilating)
Domestic Assaults	Receipt of Stolen Property
Deaths (All)	Robbery
Drunk Driving	Runaway
Embezzlement	Sex Offense, Sodomy
Extortion/Blackmail	Sex Offense, Incest
Espionage	Sex Offense, Rape
Family Offenses, Nonviolent	Sex Offenses, Pornography (Obscene Material)
Forgery	Sex Offense, Forcible
Fraud	Sex Offense, Forcible Fondling
Fraud, Welfare Fraud	Sex Offense, Sexual Assault with an object
Fraud, Wire Fraud	Sex Offense, Nonforcible
Fraud, Bad Checks	Sex Offense, Statutory Rape
Fraud, False Pretenses	Sex Offenses, Voyeurism (Peeping Tom)
Fraud, Confidence scams	Solicitation
Fraud, credit card/ATM Fraud	Spying
Fraud, Impersonation	Sodomy
False Official Statement	Theft (All types)
False Pretenses, obtaining services under ($100 or more)	Trespass of Real Property
False Swearing	Testify (Wrongful refusal)
Fleeing the scene of an accident	Unlawful/Forced Entry
Gambling	Vagrancy
Gambling Equipment Violations	Vandalism
Gambling, Betting/Wagering	Weapon Law Violations
Gambling, Operating/Promoting/Assisting	
Gambling, Sports Tampering	
Homicide	
Homicide, Murder (nonnegligent Manslaughter)	

Figure Y-1. Table of Offenses

APPENDIX Z
Uniforms and Equipment

Z.1 OVERVIEW

Members of the Ashore Navy Security Forces (NSF), while on duty, will wear the uniforms prescribed and equipment issued by the respective Security Detachment and as prescribed by Commander Navy Installations Command (CNIC). Uniforms will be worn in their entirety and in their intended fashion, and will be kept clean, well-brushed and pressed. The standard Department of Navy (DON) badge and name plate are to be displayed on the outermost garment being worn by a civilian uniformed officer. Military personnel will wear nametags as prescribed by DON regulations. Non-uniformed personnel will affix the badge to their jacket pocket or lapel, or to their belt while at a crime scene or police operation.

Buttons and other metal accessories will be clean and bright. Uniform coats/jackets must be buttoned or zippered at all times when on duty. Unauthorized objects such as sunglasses, cigarette holders, and the like may not be attached to any part of the uniform. Officers will be well groomed and present a professional appearance to maintain the public's trust and respect.

Supervisory officers (military & civilian) are charged with the responsibility of ensuring their personnel are at all times properly wearing and maintaining their uniforms and equipment. They are further tasked with overseeing conformance to established grooming standards, and that their nonuniformed subordinates alike are at all times wearing the appropriate and approved attire.

Z.2 UNIFORMS & EQUIPMENT

1. Definitions.

    a. Government equipment and/or property means and includes:

    (1) All uniforms and equipment issued to master-at-arms (MA), auxiliary security force (ASF), government service civilian employees and may include contractor support personnel.

    (2) Government-owned or leased vehicles, accessories, and government-owned property.

    (3) All other government equipment or property assigned to or assumed by a NSF member at any time.

    b. Equipment Issue and Allowances. The respective Security Detachment/Region Headquarters Antiterrorism Program Management Office (N3AT) provides all NSF personnel with uniforms or an allowance to purchase the uniforms and equipment, at no cost to the employee. Uniform allowances are provided for replacement of worn uniforms, equipment, and clothing as discussed below.

2. Uniform Allowance Government Civil Service Employees.

    a. Payment of uniform allowance shall be made to employees required to wear a uniform. Initial and replacement allowance shall be as prescribed in the most current version of Department of Defense Instruction (DODI) DoDI 1400.25, Volume 591, DoD Civilian Personnel Management System: Uniform Allowance Rates for DoD Civilian Employees.

    b. Regional Security Officers (RSO) will determine and document the appropriate amount of initial uniform allowance paid to employees based on the cost of the minimum required uniform articles as well as any additional uniform items the region deems necessary.

    c. Allowance for uniforms specifically applies to clothing which an employee is required to wear in the performance of official duties and shall be used for that purpose. Uniform allowance will be paid once a year.

Note

An employee is not required to purchase uniforms from any particular supplier as long as the uniform standards are met.

Z.2.1 Uniform Classes

The Classes of Uniforms (Civilian GS-083/085) consist of Class A Uniform—Full Dress Uniform (ceremonial) and Class B Uniform—Uniform of the Day for routine and normal assignments within the unit or Security Detachment.

1. The Class A Uniform. The class "A" uniform is worn at the direction of the Regional Security Officer or his/her designee, for ceremonial events, e.g., awards ceremonies, promotion ceremonies, and funerals. The class "A" uniform consists of the following:

    a. Long sleeve issued shirt

    b. "Sheriff" style hat

    c. Plain toe, black dress shoes

    d. Dark blue or black socks

    e. Trousers (dark, navy blue)

    f. Black belt (worn inside trouser loop)

    g. Black tie (clip-on)

    h. White gloves

    i. Badge and name plate, to be worn on outer garment

    j. Nylon Law Enforcement Duty belt and accessories.

2. Class "B" Uniform. The class B uniform is the NSF uniform of the day, as designated by the Region Commander (REGCOM) or his designee. The class B uniform consists of the summer and winter uniforms as listed below, with authorized personal items and accessories. Members of specialized units are authorized to wear supplemental uniform articles as specified in their approved standard operating procedures. An example of the Class B uniform, accessories, and equipment is found in figure Z-1.

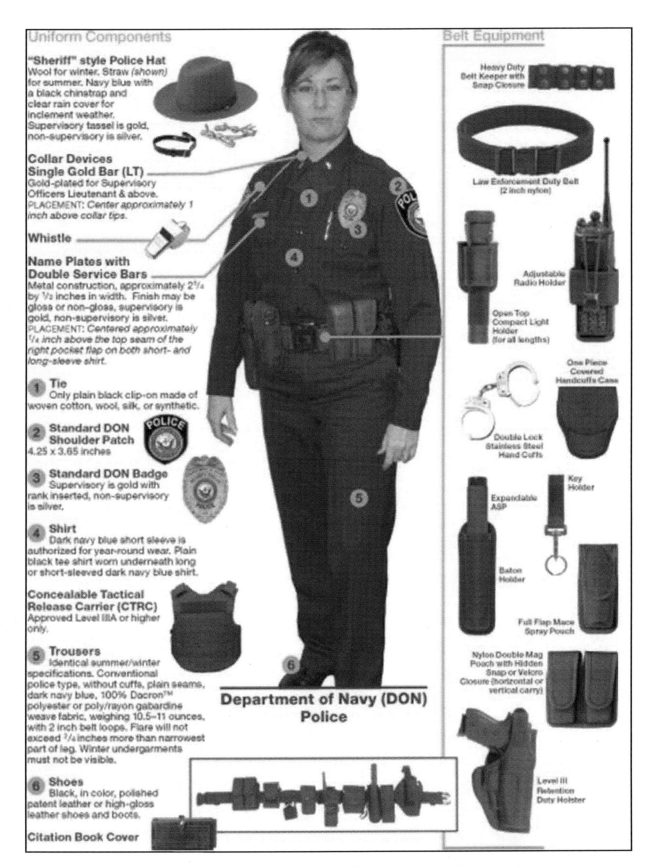

Uniform Components

"Sheriff" style Police Hat
Wool for winter. Straw *(shown)* for summer. Navy blue with a black chinstrap and clear rain cover for inclement weather. Supervisory tassel is gold, non-supervisory is silver.

Collar Devices
Single Gold Bar (LT)
Gold-plated for Supervisory Officers Lieutenant & above. PLACEMENT: *Center approximately 1 inch above collar tips.*

Whistle

Name Plates with
Double Service Bars
Metal construction, approximately 2¼ by ½ inches in width. Finish may be gloss or non-gloss, supervisory is gold, non-supervisory is silver. PLACEMENT: *Centered approximately ¼ inch above the top seam of the right pocket flap on both short- and long-sleeve shirt.*

1 Tie
Only plain black clip-on made of woven cotton, wool, silk, or synthetic.

2 Standard DON
Shoulder Patch
4.25 x 3.65 inches

3 Standard DON Badge
Supervisory is gold with rank inserted, non-supervisory is silver.

4 Shirt
Dark navy blue short sleeve is authorized for year-round wear. Plain black tee shirt worn underneath long or short-sleeved dark navy blue shirt.

Concealable Tactical
Release Carrier (CTRC)
Approved Level IIIA or higher only.

5 Trousers
Identical summer/winter specifications. Conventional police type, without cuffs, plain seams, dark navy blue, 100% Dacron™ polyester or poly/rayon gabardine weave fabric, weighing 10.5–11 ounces, with 2 inch belt loops. Flare will not exceed ¾ inches more than narrowest part of leg. Winter undergarments must not be visible.

6 Shoes
Black, in color, polished patent leather or high-gloss leather shoes and boots.

Citation Book Cover

Department of Navy (DON)
Police

Belt Equipment

Heavy Duty Belt Keeper with Snap Closure

Law Enforcement Duty Belt (2 inch nylon)

Adjustable Radio Holder

Open Top Compact Light Holder (for all lengths)

One Piece Covered Handcuffs Case

Double Lock Stainless Steel Hand Cuffs

Key Holder

Expandable ASP

Baton Holder

Full Flap Mace Spray Pouch

Nylon Double Mag Pouch with Hidden Snap or Velcro Closure (horizontal or vertical carry)

Level III Retention Duty Holster

Figure Z-1. Department of the Navy Police Lieutenant and Accessories

a. (Summer uniform) Male/Female

    (1) Black shoes/boots

    (2) Dark blue or black socks

    (3) Trousers (dark, navy blue in color)

    (4) Black belt (worn inside trouser loop)

    (5) Short sleeve shirt (dark, navy blue/white for supervisors)

    (6) Dark blue T-shirt

    (7) Hat (Sheriff style felt/straw optional for nonsupervisory personnel/required for supervisors)

    (8) Ball cap-style hats (dark, navy blue with Police or Guard white lettering for nonsupervisors)

    (9) Law Enforcement Duty belt and accessories

  (10) Standard DON Badge and Name Plate

  (11) Insignia of rank

  (12) Commendations/Shooting Awards (optional)

  (13) Jacket (convertible type, navy blue)

  (14) Windbreaker (dark, navy blue).

b. (Winter Uniform) Male/Female

    (1) Black shoes/boots

    (2) Dark blue or black socks

    (3) Trousers (dark, navy blue winter)

    (4) Black belt (worn inside trouser loop)

    (5) Black tie (clip-on)

    (6) Sheriff style hat (Felt)

    (7) Long sleeve shirt (dark, navy blue/white for supervisors)

    (8) Law enforcement duty belt and accessories

    (9) Standard DON Badge/Name Plate (worn on outer garment)

  (10) Insignia of rank

  (11) Winter coat

  (12) Approved scarf (dark blue/black) (optional)

  (13) Black gloves

  (14) Commendations/Shooting Awards (optional).

Z.2.2 Uniforms and Equipment Required While On Duty

In addition to the uniforms, NSF uniformed officers must be equipped with the following items while on duty, unless excused by their Installation Security Officer (SO), Installation Commanding Officer (CO) or Regional Security Officer (RSO).

Z.2.2.1 Uniformed Officers (GS-083/085)

1. 2" Nylon law enforcement duty belt

2. Level III retention duty holster

3. Double-locking high security handcuffs, key and nylon case

4. Baton (ASP) and holder: Authorized impact nonlethal weapon

5. Oleoresin Capsicum (OC) spray and nylon pouch

6. Flashlight and nylon open top compact flashlight holder

7. Portable radio (charged & operational) w/nylon adjust radio holder

8. Nylon duty pouch (latex & search gloves, medical pouch)

9. Nylon key holder

10. Heavy duty nylon belt keepers (4 pack)

11. 9mm Beretta semiautomatic pistol

12. Two magazine clips

13. Pistol lanyard

14. "H" Harness (optional)

15. Level IIIA Body Armor System

16. Whistle (silver nonsupervisor/gold supervisory)

17. Name Plate (silver nonsupervisor/gold supervisory)

18. Standard DON Badge (silver nonsupervisor/gold supervisory)

19. Valid state driver's license

20. Certification cards (e.g., weapons, CPR, First-Aid)

21. Lightweight helmet (LWH)

22. Personal protective equipment (PPE) IAW OPNAVINST 3440.17, Navy Installation Emergency Management (EM) Program

23. Citation book/motor vehicle code book

24. Notebook and pen

25. Equipment bag.

Supervisors will inspect their personnel daily at roll call formation to ensure uniforms and other equipment are clean and in good order. Supervisors will also ensure each officer is properly equipped and attired, and that he/she is physically fit for duty.

Z.2.2.2 Navy Master-at-Arms (MA) & Auxiliary Security Force (ASF)

In addition to the uniforms, military personnel are guided by Navy Uniform Regulations in wearing and maintenance of utility uniforms identified for Security duties.

1. 2" Nylon law enforcement duty belt (NSF/ASF)

2. Level III retention duty holster (NSF/ASF)

3. Double-locking high security handcuffs, key and nylon case (NSF/ASF)

4. Baton (ASP) and holder: Authorized impact nonlethal weapon (NSF/ASF)

5. Oleoresin Capsicum (OC) spray and nylon pouch (NSF/ASF)

6. Flashlight and nylon open top compact flashlight holder (NSF/ASF)

7. Portable radio (charged & operational) and nylon adjustable radio holder (NSF/ASF)

8. Nylon duty pouch (latex & search gloves, medical pouch) (NSF/ASF)

9. Nylon key holder (NSF/ASF)

10. Heavy duty nylon belt keepers (4 pack) (NSF/ASF)

11. 9mm Beretta semiautomatic pistol (fully loaded) (NSF/ASF)

12. Two magazine clips (NSF/ASF)

13. Pistol lanyard (NSF/ASF)

14. "H" harness (NSF) (optional)

15. Level IIIA Body Armor System (issued) (NSF/ASF)

16. Whistle (NSF/ASF)

17. Standard DON badge (NSF)

18. Valid state driver's license (NSF/ASF)

19. Notebook and pen (NSF/ASF)

20. Certification cards (e.g., weapons, CPR/First-Aid) (NSF/ASF)

21. Lightweight helmet (LWH) (NSF/ASF)

22. Personal protective equipment (PPE) (NSF/ASF).

23. Citation book (NSF)

24. Motor vehicle code book (NSF).

Z.2.2.3 Plainclothes Investigators

Investigators are to be dressed in business casual attire <u>unless</u> assigned in an undercover capacity. Business casual attire <u>does not</u> include the wearing of sneakers, jeans, shorts or shirts without collars. Investigators appearing in court shall be dressed in business attire. When an investigator is on duty in civilian dress he/she must be equipped with the following unless assigned to an undercover/covert unit or excused by his/her supervisor:

1. 1–1 1/2" belt (inside belt loop)

2. Double-locking high security handcuffs, key and case (figure Z-2)

3. Duty holster, level III retention duty holster (figure Z-3)

4. Baton (ASP) and holder: Authorized impact nonlethal weapon

5. OC spray and pouch

6. Flashlight and holder

7. Portable radio (charged & operational)

8. 9mm Beretta semiautomatic pistol (fully loaded) (figure Z-2)

9. Two magazine clips (figure Z-4)

10. Type IIIA body armor system (concealable system)

11. Standard DON badge/investigator credentials

12. Valid state driver's license

13. Certification cards (e.g., weapons, CPR)

14. Notebook and pen

15. Binoculars.

Figure Z-2. Sidearm/Handcuffs

Figure Z-3. Cell w/Magazine

Z.3 COURT APPEARANCES

Official Business. Officers will wear the uniform of the day when appearing in court on official government business (local policy enforced on carrying government issued firearms). The uniform, whenever worn, will be neat and clean. Off-duty officers may wear civilian attire, provided the male officer is dressed in a business suit with tie, or sports jacket with tie, and the female officer is dressed in appropriate daytime business apparel (local policy enforced).

Personal Business. Members of the department who attend court on personal business are prohibited from carrying any weapon and wearing the departmental uniform into any courthouse or court annex where the case is to be adjudicated.

Z.4 BODY ARMOR PROTECTION

Z.4.1 Body Armor

Body armor is an item of personal protective equipment that provides protection against specific ballistic threats within its coverage area (See figures Z-4 through Z-10). Body armor must meet current NIJ standards. This standard does not apply to helmet systems, eye protection systems, or other transparent armor systems worn on the body. The latest information regarding approved body armor protection can be found at http://dcfpnavymil.org/.

Body armor saves lives. According to industry estimates, more than 1,200 police officers in the United States have been saved from death or serious injury because they were wearing their armor. This figure includes lives saved and serious injury avoided from assaults and accidents as well as from ballistic threats.

Z.4.2 Soft Armor

Soft armor refers to types consisting of multiple plies of ballistic fibers either woven or nonwoven. Soft armors are designed to defeat relatively low energy (less than 1,000 ft·lb) soft or blunt projectiles including steel fragments and lead handgun projectiles.

Z.4.3 Hard Armor

Hard armor refers to types including fiber-reinforced plastics (FRPs), metals (for example, aluminum, titanium, and steel), ceramics (for example, aluminum oxide, silicon carbide, and boron carbide), and composites of these various types of materials.

Z.4.4 Insert

A removable or nonremovable unit of ballistic material that can enhance the ballistic performance of the armor panel in a localized area but not over the entire area intended for ballistic protection. Some inserts are known as trauma packs, trauma plates, or trauma inserts, but other forms of inserts are possible.

Z.4.5 Inserts, Plate

Hard armor plates or semi-rigid plates are intended to be inserted into pockets of flexible vests and jackets to provide increased protection, particularly to provide protection against rifle threats.

Z.4.6 Service Life

Unless otherwise stated, soft armor systems shall have a minimum service life of five years of continuous use in all types of typical field use if not impacted by ballistic projectiles and 15 years including intermittent storage periods ranging from one month to five years, maximum duration. The system must be cleaned before intermittent storage period. Due to the nature of most hard armor plates accurate estimates of service life during continuous field use are not readily available. Steps should be taken to ensure the longest service life possible.

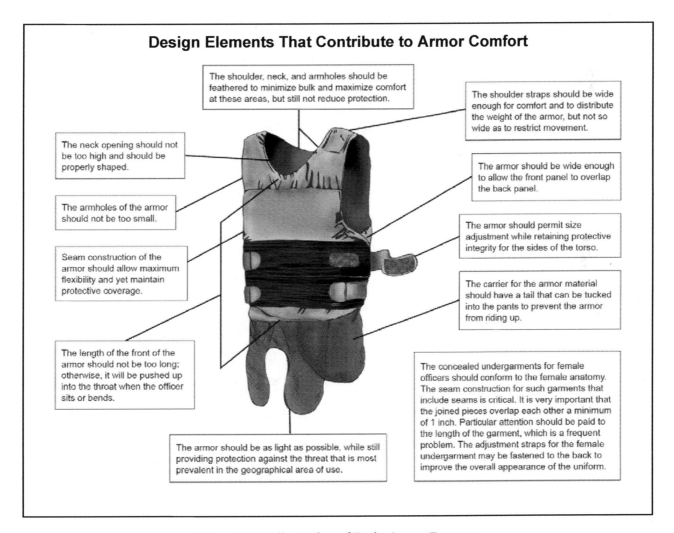

Figure Z-4. Illustration of Body Armor Features

ITEM	DESCRIPTION	NSN/PART NUMBER
MODULAR TACTICAL VEST (MTV)	MTV, X-SMALL	8470-01-547-5919
	MTV, SMALL	8470-01-547-5950
	MTV, MEDIUM	8470-01-547-5938
	MTV, LARGE	8470-01-547-5949
	MTV, X-LARGE	8470-01-547-5954
LIFE PRESERVER, VEST	LBT-2564A TACTICAL FLOTATION VEST	4220-01-539-7252
BODY ARMOR, CONCEALABLE	MALE, SMALL	8470-01-474-5127
	FEMALE, SMALL	8470-01-474-5135
	MALE, MEDIUM	8470-01-474-5140
	FEMALE, MEDIUM	8470-01-474-5143
	MALE, LARGE	8470-01-474-5146
	FEMALE, LARGE	8470-01-474-5148
	MALE, X-LARGE	8470-01-474-5151
	FEMALE, X-LARGE	8470-01-474-5152
	MALE, XX-LARGE	8470-01-521-0905
	FEMALE, XX-LARGE	8470-01-521-0907

Figure Z-5. Body Armor Systems/Accessories Approved for Use by Navy Security Forces

Features
- Side Opening
- Six Point Removable Strapping System
- Micro-Fiber Carrier
- Breathable Anti-microbial Mesh Inner Liner
- Front and Back Shirt Tails
- Front Loading Easy Access Ballistic Enclosure

Options
- Steel Blunt Trauma Insert
- Soft Blunt Trauma Insert
- Tactical Response Carrier (TRC)
- NIJ Certified Level II or IIIA Heat-Sealed Ballistic Panels

Figure Z-6. Elite Tactical Carrier Level IIIA

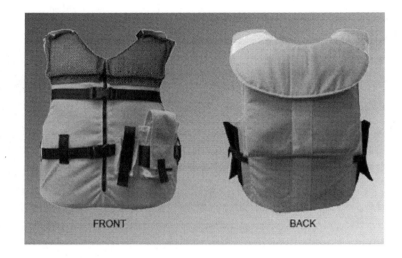

	Features:
	• Side Opening Design
	• 500 Denier Cordura Outer
	• Internal Cummerbund
	• Adjustable Shoulder and Side
	• Standard Radio, and Utility Pouches
	• Scotchlite Reflective Material on Back and Collar
	• Front Loading Ballistic Panel
	• NIJ Certified Level II and IIIA Ballistic Packages
	• Heat Sealed Ballistics and Floatation Material
	• Optional Colors Available
	• Optional Groin Strap

Figure Z-7. Port Authority Vest (Maritime) Level IIIA

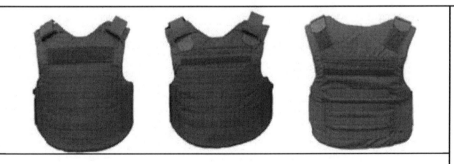

CTR is designed as a low profile ballistic carrier to provide both concealable and outer tactical applications. The CTR is equipped with both front and back top load plate pockets which can be manufactured to fit both military and law enforcement specified plates. The CTR photos in figure Z-8 show both heavy duty and low profile strapping systems along with optional grid-lock webbing attachment system.

Features:
- Five hundred Denier Nylon Cordura
- Optional Grid-lock Webbing (Front, Back, Sides)
- Six Point Strapping System
- Internal Cummerbund for secure fit
- Hook and Loop side closures with additional Strapping
- Front and Back Top Load Hard Armor Plate Pockets (Manufactured to fit plates)
- Available in optional colors (Black, Desert Camo, Navy, Woodland Camo)

Figure Z-8. Concealable Tactical Response (CTR)

 	Kit includes: • Concealable armor (worn under shirt) • Tactical outer carrier • Trauma pad inserts • Shoulder armor pads • Kit bag Additional items available: • Lower back and groin plates • Bicep pads • Collar pads	The classification of an armor panel that provides two or more levels of NIJ ballistic protection at different locations on the ballistic panel shall be that of the minimum ballistic protection provided at any location on the panel.

Figure Z-9. Tactical Domestic Agent Ballistic Vest Kit

	Carrier Options: • Side Plate/Ballistic Pouches (Mandatory side Grid-lock Webbing on Sides) • Grid-lock Webbing (front and back) • Assorted Tactical Pouches	Side Pouches designed to accommodate hard armor plates as well as soft ballistic panels for additional protection.

Figure Z-10. Side Pouch

Z.4.7 Shelf Life

The depot storage life shall meet the requirements specified in the acquisition document. The system shall resist any structural and ballistic degradation due to depot storage. Depot storage may include exposure to temperature extremes.

Z.5 MINIMUM PROTECTION STANDARDS (LEVEL IIIA/LEVEL IV)

The approved standard for the Ashore enterprise is a Level IIIA system capable of accepting ESAPI and/or a Level IV system. Level IIIA body armor provides the highest level of protection available in concealable body armor and provides protection from high velocity 9mm and .44 Magnum rounds of ammunition.

Level IIIA armor is suitable for routine wear in many situations; however, NSF units located in hot, humid climates should continually and carefully monitor their Level IIIA body armor systems for any signs of performance degradation. Level IV armor protects against high powered rifle rounds and is clearly intended for use only in tactical situations when the threat warrants such protection.

The ballistic threat posed by a bullet depends, among other things, on its composition, shape, caliber, mass, angle of incidence, and impact velocity. Because of the wide variety of bullets and cartridges available in a given caliber and because of the existence of hand-loaded ammunition, armors that will defeat a standard test bullet may not defeat other loadings in the same caliber.

For example, an armor that prevents complete penetration by a .40 Smith and Wesson (S&W) test bullet may or may not defeat a .40 S&W bullet with higher velocity. In general, an armor that defeats a given lead bullet may not resist complete penetration by other bullets of the same caliber of different construction or configuration. The test ammunition specified in this standard represents general, common threats to law enforcement officers.

Z.5.1 Level IIIA (High Velocity 9mm; .44 Magnum)

1. This armor protects against 9mm Full Metal Jacketed Round Nose (FMJ RN) bullets, with nominal masses of 8.0 grams (g) (124 grain (gr)) impacting at a minimum velocity of 427 meters per second (m/s) (1400 feet per second (ft/s)) or less, and .44 Magnum Semi-Jacketed Hollow Point (SJHP) bullets, with nominal masses of 15.6 g (240 gr) impacting at a minimum velocity of 427 m/s (1400 ft/s) or less. It also provides protection against most handgun threats.

2. Provides multiple-hit protection against .44 Magnum lead Semi-Wad Cutter gas checked bullets and 9mm FMJ bullets.

Z.5.2 Level IV (Armor Piercing Rifle)

This armor protects against .30 caliber armor piercing (AP) bullets (U.S. Military designation M2 AP), with nominal masses of 10.8 g (166 gr) impacting at a minimum velocity of 869 m/s (2850 ft/s) or less. It also provides at least single hit protection against .30–06 AP bullets.

Z.6 FIELDED NAVY BODY ARMOR SYSTEMS

The BA systems listed below represent body armor systems authorized for use by U.S. Navy activities and prescribed for use on Navy installations. This listing is subject to update in technology and National Institute of Justice (NIJ) systems review.

The criteria utilized for authorization includes verification of performance specification as described in purchase descriptions identified in references above and NIJ 0101.06 Ballistic Resistance of Body Armor standard certification. Purchases of armor systems based on U.S. Army and U.S. Marine Corps specifications shall be accompanied by First Article Testing and Lot Acceptance Testing documentation.

Purchase of other soft armor systems shall be accompanied with NIJ 0101.06 Level IIIA certification, IOTV fragmentation protection documentation and documentation of a backface deformation of less than 43mm when used with an ESAPI plate shot with 7.62x54R Type LPS at 2,750 + 50 feet per second. All documentation shall be forwarded to the Body Armor Technical Warrant Holder for validation prior to purchase.

This listing is exclusive to U.S. Navy personnel assigned to law enforcement units (e.g., Master-at-Arms and DON police and security guards). CNIC Navy Security Forces shall use NIJ 0101.06 Level IIIA armor systems and should be capable of accepting ESAPI plates. Law enforcement units shall verify all body armor systems in use meet NIJ standards and retain a copy of the "Notice of Compliance" for the particular body armor system for the service life of the system and provide a copy of the same to the Body Armor Technical Warrant Holder.

Z.7 APPROVED BODY ARMOR SYSTEMS (GSA AVAILABLE)

Training. A use and care manual accompanies each body armor system listed below and explains with text and drawings how to adjust, wear and maintain the system.

Z.8 HEAD PROTECTION

Z.8.1 Lightweight Helmet (LWH)

The LWH replaces the Personal Armor System for Ground Troops (PASGT) helmet, also known as the Kevlar. The LWH will provide protection in various operating environments from fragmentation and 9mm small arms projectiles (figures Z-11 and Z-12). The LWH is issued primarily to CONUS installations for use during riot control operations, emergency response team operations and certain MWD support requirements.

Figure Z-11. Light Weight Helmet

Figure Z-12. Inside View of Light Weight Helmet

1. Features. The LWH remains the PASGT design but makes use of lighter materials made available by new technologies. The LWH is approximately 3 to 6 ounces lighter than the PASGT depending on helmet size. It provides greater combat effectiveness through greater comfort and improved fit. It features a retention system and pads that will reduce stress and fatigue on the neck and shoulders. It is available in five sizes ranging from X-small to X-large.

2. Components

   a. Ballistic Shell

   b. Suspension Hook Disks (Velcro)

   c. Suspension Pads (3/4 or 1-inch thick)

   d. Retention Assembly

   e. Attaching Hardware

   f. Edging, Finish, and Labels

   g. Use and Care Manual.

3. Training. A use and care manual accompanies each helmet and explains with text and drawings how to adjust, wear, and maintain the helmet.

4. Supply/Logistics. Defense Logistics Agency (DLA) (DSCP) has been designated as the sustaining organization for this item per fielding message 051846Z Nov 2003.

TAMCN: C1321

ID No: 08744B

Production—The Gentex Corporation in Carbondale, PA produces the LWH.

Manuals:

User Logistic Support Summary (ULSS): 08744-15, Lightweight Helmet (LWH) (PCN: 13200874400), Located at https://pubs.ala.usmc.mil/pcnsearch.asp

Technical Manual: TM 08744B-12&P Operator's Care and Use Manual for Lightweight Helmet (LWH) (PCN: 50008744000), Located at https://pubs.ala.usmc.mil/pcnsearch.asp

Z.8.2 Advanced Combat Helmet (ACH)

The ACH is normally issued to OCONUS areas where the possibility of combat operations exist. The ACH is a modular system that weighs less, fits better, and is more comfortable than its predecessor (figures Z-13 and Z-14). Modular, flame-retardant, and moisture-resistant pads act as the suspension system between the wearer's head and the helmet. The cotton/polyester chin strap, a four-point design, allows for quick adjustment and includes a new Ballistic Protective Pad for the neck that adds ballistic protection between the bottom of the helmet shell and the top of the Interceptor Body Armor collar. The edge of the ACH shell is finished with rubber trim. The ACH is available in five sizes from small through XX-large.

Figure Z-13. Advanced Combat Helmet (ACH)

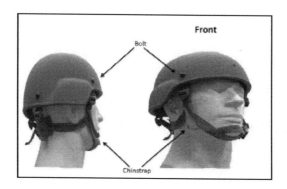

Figure Z-14. ACH

1. Component Materials

   a. Helmet Shell: Aramid fabric. Chinstrap and Retention System: cotton/polyester webbing and foam nape pad; mounting hardware; and a combination of metal and composite materials.

   b. Color: Helmet shell: currently Foliage Green 504. Chinstrap/Buckles: currently black, and combined Green 483/Tan—Camouflage Cover: Universal Camouflage Pattern.

   c. System Weight: Small: 2.93 lbs; Medium: 3.6 lbs; Large: 3.31 lbs; X-Large: 3.60 lbs; and XX-large: 3.77 lbs.

   d. Size: Currently available in five shell sizes (small, medium, large, X-large, and XX-large).

2. Training

   a. A use and care manual accompanies each helmet and explains with text and drawings how to adjust, wear, and maintain the helmet.

   b. Program Executive Office (PEO) Soldier is the Army acquisition agency responsible for everything a soldier wears or carries. View our website at https://peosoldier.army.mil for more information on PEO Soldier.

Z.9 INDIVIDUAL AND PROTECTIVE EQUIPMENT

Individual and protective equipment is identified in figure Z-15.

ITEM	DESCRIPTION	NSN/PART NUMBER	GSA AVAILABLE
2" Ultra Duty Belt with Hook and Loop Lining	Double layer of tough 2" nylon web with a polymer stiffener for support for the holstered gun and other belt gear.		YES
Level III Retention Duty Holster	Top-draw holster with standard ride. Three separate retaining devices.		YES
Baton Holder	Dual belt loop design accommodates both 2" (50mm) and 2.25" (58mm) belts. Covered hole in back of pouch allows baton to be placed in pouch fully expanded.	Patrol Tek 8012 Expandable Baton Holder	YES

Figure Z-15. Individual and Protective Equipment Available for Navy Security Force (Sheet 1 of 4)

ITEM	DESCRIPTION	NSN/PART NUMBER	GSA AVAILABLE
ASP BLACK CHROME TACTICAL BATONS	Baton is durable, easier to handle and provides excellent retention in all conditions. 26" is the authorized ASP.	9B 8465-01-572-8488	YES
Handcuffs w/Case	High Security Chain-Linked Nickel Handcuffs w/Push Pin Double Locking System. Single Cuff Case with Key Slot. Dual belt loop fits 2" and 2.25" width duty belts.	Model 104P PatrolTek 8001 Single Cuff Case with Key Slot	YES
Belt Keepers–4 pack)	Heavy Duty Nylon Belt Keeper—Features a snap closure, soft lining to protect gear—Hand washable, fade, scratch and weatherproof resistant.	9B 8315-01-446-8613	YES
OC Spray Pouch	Hidden snap closure—Full flap helps secure canister—Web belt loop fits 2" (50mm)–2.25" (58mm) duty belts.	9B 8465-01-572-6687	YES
OC Spray	Nonlethal—Nonflammable aerosol defense spray—Heavy stream 4.0 oz. 2-year shelf life.	Oleoresin Capsicum (OC) Formula	YES

Figure Z-15. Individual and Protective Equipment Available for Navy Security Force (Sheet 2 of 4)

ITEM	DESCRIPTION	NSN/PART NUMBER	GSA AVAILABLE
Flashlight Holder	Open Top Compact Light Holder. 2 sizes fit most 1" diameter lights. Fits 2 1/4" web belt loop. Holds all flashlight lengths.	AccuMold 7326	YES
Rechargeable Flashlight	Improved corrosion resistance and durability. Water- and shock-resistant. Battery recharges up to 1000 times.	MagLite MagCharger Rechargeable Flashlight *Item#* FL070	YES
Key Holder	Simple and effective. Fits 2-1/4" belt. Nylon web construction ingle.	AccuMold 6405 Key Holder	YES
Magazine Pouch	Horizontal or vertical carry. Four sizes to fit most magazines. Injection molded belt loop. For use with 2.25" duty belts.	9B 8465-01-570-9849	YES
Adjustable Radio Holder	Holds compact handheld radios. Adjustable	AccuMold 7323 Adjustable Radio Holder	YES

Figure Z-15. Individual and Protective Equipment Available for Navy Security Force (Sheet 3 of 4)

ITEM	DESCRIPTION	NSN/PART NUMBER	GSA AVAILABLE
Citation Book Cover	Holds one citation book and has one storage pocket.	580 Single Citation Book Cover	YES
Camelbak Drinking System	Camelbak drinking system. An inexpensive, lightweight drinking system with replaceable bladders. Black, desert, woodland 2 liter capacity	8465-01-517-2165	YES
Equipment Bag	Industry standard equipment bag	Blackhawk Police Equipment Bag	YES

Figure Z-15. Individual and Protective Equipment Available for Navy Security Force (Sheet 4 of 4)

Z.9.1 Duty Holster

Z.9.2 Understanding the Levels of Retention for Duty Holsters

Levels of Retention for Duty Holsters should not be confused with small arms condition levels e.g., (conditions 4, 3, & 1).

Note

The National Institute of Justice (NIJ) is considered as the national authority on duty holsters and other protective equipment for law enforcement personnel.

NIJ establishes minimum requirements and test methods for duty holsters designed to provide law enforcement personnel with the ability to securely carry, easily deploy and resecure their duty weapon. NIJ standards addresses gun-retention capability of duty holsters. The standard will define two classifications of holsters based on the style of retention: manual and automatic. Level III holsters are the minimum standard for use by installation NSF personnel (Figure Z-16).

The retention level of the holster refers to the number of retention devices the user must release or move the gun past in order to draw the pistol from the holster. A Level III holster, as shown below is one where three separate retaining devices.

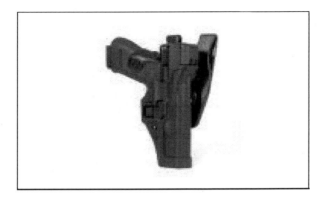

Figure Z-16. Level III Holster

1. The release must be integral with the draw. Releasing the retention device/s present should be an integral part of the drawing process. Any retention device in use should release at the same time as the grip on the stocks of the pistol is taken.

2. Gripping the firearm. The hand must be able to take a full firing grip on the stocks of the weapon while it is still in the holster. A well-designed holster should allow the firing hand to fully grip the stocks of the firearm in the normal firing grip while the firearm is still in the holster. This is central to obtaining accurate shot placement. No portion of the holster body should interfere in any way with the drawing process.

3. Locking the holster to the belt. The holster should come equipped with some form of device that locks the holster rigidly to the belt and prevents it from moving. This is an inherent problem for military personnel wearing camouflage uniforms since their rigs are secured over the blouses of the uniform. Duty holsters generally are bumped several times a day on chairs, entering or exiting cars or on furniture. Despite this, they should not move. If the holster shifts on the belt during the draw, the technique is compromised.

4. Weapons Retention Training. Training in weapons retention must be mandatory and encouraged as often as possible for familiarization purposes. There is a firearm present at every situation an officer attends–his or her own gun. Any holster can be defeated and some of the worst real life horror stories revolve around officers having their own firearm taken from them and then being shot with the weapon.

Z.10 VEHICLE EQUIPMENT

Z.10.1 Law Enforcement and Physical Security Vehicles

Vehicles used for LE/PS functions shall be of the minimum body size, maximum fuel efficiency, and lowest overall cost to meet the mission.

Z.10.2 Police Patrol Vehicles

Police patrol vehicles are radio equipped emergency-configured vehicles. Patrol vehicles are utilized for 24x7 patrol shift work. The number of patrol vehicles for an activity will be guided by the number of validated patrol posts. Patrol post validation is a function of CNIC (N3AT). For activities where police patrol vehicles endure high mileage and/or high operating hours, patrol vehicles shall be midsize law enforcement sedans. Law enforcement configured sedans feature heavy duty electrical systems and wiring harnesses for police equipment. They also feature heavy duty cooling systems for long operating hours at low speeds. For small activities with short patrol distances and lower vehicle operating hours, generic (not law enforcement package) midsize sedans may be considered. In either case, a midsize sedan is the largest sedan that will be authorized for Navy police patrol use. When assessing the vehicle type for patrol use at a particular activity, the terrain and weather should be considered. Where the terrain of validated patrol posts dictates higher ground clearance vehicles and/or off road conditions, the patrol vehicle fleet may include a limited number of four wheel drive SUVs or light trucks.

Similarly, in areas where heavy snowfall is routine, a limited number of four wheel drive SUVs should be considered. For activities where a validated patrol post can only be patrolled by an all terrain vehicle (ATV), use of ATVs is appropriate and the number of patrol vehicles shall be reduced accordingly on a one for one basis. ATVs are otherwise not authorized due to safety concerns. Motorcycles are not authorized for Navy law enforcement patrols.

Z.10.3 Vehicle Equipment

To the maximum extent possible, patrol vehicles shall be outfitted in a way that minimizes modifications to the vehicle. Such modifications add to the turn-in cost of a vehicle at the end of its service. Nontactical vehicles permanently assigned as NSF patrol vehicles should be equipped with the following:

1. Warning Light System. Use an Exterior Roof Mounted Super LED Multicolor (Red/Blue/Amber) Light Bar; front LED head light flashers (clear); front grill LED lights (red/blue); rear deck LED lights (red/blue) and rear tail light LED flashers (clear).

2. Siren System. Minimum 100 Watt Siren Speaker w/internal PA system. Mount the siren in concert with the all warning light systems.

3. Spotlight. Use either a portable spotlight (battery-operated or cigarette lighter plug-in type) or a permanently mounted spotlight. If you use a permanent-mount type, mount the spotlight on the vehicle warning light system.

4. Land Mobile Radio. Each patrol vehicle shall be equipped with a two way radio. As Land Mobile Radios (LMR) are fielded, this will be the radio of choice. In the interim, two way radios should be of a type that is interoperable with existing base radio systems. Radios shall be installed so the driver has easy access to all radio controls and microphones and so radio equipment does not interfere with safe vehicle operation.

5. First Aid Kit. A general purpose first aid kit (FSN 6545-00-922-1200 or equivalent).

6. Extinguisher. 2 1/2 pound ABC rated dry chemical fire extinguisher.

7. Bloodborne Pathogen Protective Kit (required in all installation security department response vehicles). Contents must include one-way respiratory cardio-pulmonary resuscitation (CPR) mask, surgical gloves, eye protective goggles or glasses with side shields, surgical mask, and surgical gown.

8. Central control console with power ports, flip-up arm rest and dual cup holders.

9. Mobile Data Terminal. Each patrol vehicle shall be equipped with a mobile data terminal (MDT) as these systems are fielded. MDT shall be installed to allow the driver unencumbered access to all radio and emergency equipment controls. The MDT is not to be used when the vehicle is moving for safety reasons.

All patrol vehicles will also have:

1. Full partition cage with sliding windows and space saver insert partitions. Not all police patrol vehicles must be configured with prisoner transport shields. Where prisoner transport is an infrequent requirement, a limited number of vehicles shall be so equipped. Where possible, prefabricated shields that fit the vehicle body shall be procured. When necessary, in the absence of a prefabricated shield, custom fabricated shields are authorized.

2. Rear door and window steel bars.

3. Rear windows and rear doors dark tint.

4. Electronic lockable dual gun rack.

Z.10.4 Vehicle Identification Markings

All installation security vehicles should be marked as identified in figures Z-17 and Z-18 below. Reflective decal graphics are required when marking security vehicles. Only the type and style as identified in figures Z-18 and Z-19 are recommended for installation security departments.

Z.10.5 Vehicle Installed Video Surveillance Systems

Vehicle installed video surveillance systems may be used as a tool to enhance officer safety and the effectiveness of audio/video evidence supporting the NSF law enforcement and public safety role. The use of vehicle installed video cameras provides excellent evidentiary support for law enforcement. Region or installation policy must be developed to govern the use of video and audio evidence and disposition of video and audio data prior to equipment purchase. Vehicle Installed video surveillance systems purchased for use by installation security departments must provide digital audio and video capability as well as forward and rearward video and audio capability. The vehicle mounted video system may only be used when actively engaged in patrol activities and cannot be in use when the vehicle is parked and unattended. State and local laws should be considered prior to purchase.

Z.10.6 Speedometer Calibration

Calibrate the speedometers of all traffic patrol vehicles at least semiannually or sooner if local laws are more stringent. Recalibrate the vehicle's speedometer any time there are major maintenance repairs to a traffic patrol vehicle's transmission, differential, speedometer, or after tire replacement.

Z.10.7 Vehicle and Vehicle Equipment Care

Before each tour of duty, inspect vehicles and vehicle equipment for safety and maintenance deficiencies. Test all warning lights, sirens, public address systems, spotlights, etc. Report deficiencies as soon as possible. Vehicles should always present a clean appearance, weather permitting.

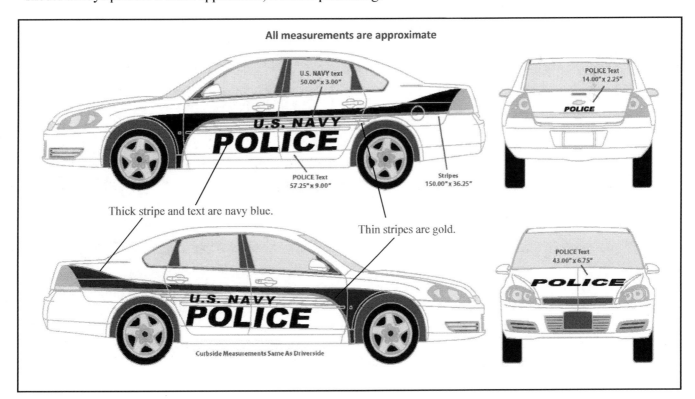

Figure Z-17. Example of a U.S. Navy Police Sedan

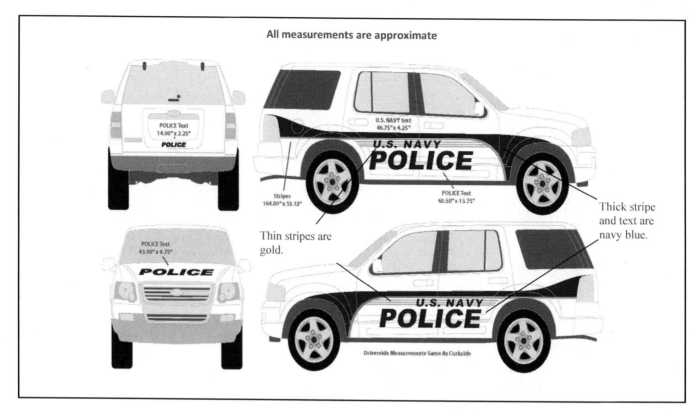

Figure Z-18. Example of a U.S. Navy Police Sport Utility Vehicle

Z.10.8 Military Working Dog (MWD) Team Vehicles

To the maximum extent possible, two wheel drive midsize SUVs will be the typical MWD team vehicle. Where the terrain dictates higher ground clearance vehicles and/or off road conditions, a limited number of four wheel drive SUVs may be leased. MWD team vehicles shall be equipped with vinyl flooring and cages in their rear cargo area. Vehicle will be clearly marked "Military Working Dog" with removable/magnetic decals as appropriate.

Z.10.9 Police Administrative Vehicles

Nonpatrol vehicles are not to be outfitted as emergency response vehicles, patrol cars, or mobile command posts. It is recommended that hybrid, compact or subcompact vehicles be used for administrative vehicles. Where low speed vehicles (mules, golf carts, GEM Cars, etc.) can be utilized, their use is authorized and encouraged.

REFERENCES

GENERAL

Chemical Weapons Convention

Federal Bureau of Investigation (FBI) and Department of Justice (DOJ) Instruction, Guidelines for Preparation of CJIS Division Fingerprint Cards

Homeland Security Presidential Directive-5 (HSPD-5), Management of Domestic Incidents

Manual for Courts-Martial (MCM), United States (2008 Edition)

Title 10, U.S.C., Armed Forces

Title 18, U.S.C., Crimes and Criminal Procedure

UFC 4-010-01, DoD Minimum Antiterrorism Standards for Buildings

UFC 4-022-01, Security Engineering: Entry Control Facilities/Access Control Points

UFC 4-022-03, Security Engineering: Fences, Gates, and Guard Facilities

DOD AND JOINT PUBLICATIONS

CJCSI 3121.01 (series), Standing Rules of Engagement/Standing Rules for the Use of Force for U.S. Forces (U)

DOD O-2000.12-H, DoD Antiterrorism Handbook (FOUO)

DODD 1030.01, Victim and Witness Assistance

DODD 2000.12, DoD Antiterrorism (AT) Program

DODD 3000.3, Policy for Non-Lethal Weapons

DODD 5200.27, Acquisition of Information Concerning Persons and Organizations not Affiliated with the Department of Defense

DODD 5210.56, Carrying of Firearms and the Use of Force by DoD Personnel Engaged in Security, Law and Order, or Counterintelligence Activities

DODD 5525.4, Enforcement of the State Traffic Laws on DoD Installations

DODD 6400.1, Family Advocacy Program (FAP)

DODI 1030.2, Victim and Witness Assistance Procedures

DODI 2000.16, DoD Antiterrorism (AT) Standards

DODI 5505.11, Fingerprint Card and Final Disposition Report Submission Requirements

JP 1-02, Department of Defense Dictionary of Military and Associated Terms (as amended)

MULTISERVICE PUBLICATIONS

AR 190-5/OPNAV 11200.5D/AFI 31-218(I)/MCO 5110.1D/DLAR 5720.1, Motor Vehicle Traffic Supervision

FM 3-11.21/MCRP 3-37.2C/NTTP 3-11.24/AFTTP(I) 3-2.37, Multiservice Tactics, Techniques, and Procedures for Chemical, Biological, Radiological, and Nuclear Consequence Management Operations

FM 3-22.40/MCWP 3-15.8/NTTP 3-07.3.2/AFTTP(I) 3-2.45/USCG Pub 3-07.31, Tactical Employment of Nonlethal Weapons (NLW)

NTTP 3-20.6.29M/COMDTINST M3120.18/MCWP 3-35.8, Tactical Boat Operations

NAVY PUBLICATIONS

BUMEDINST 6120.20 (series), Competence for Duty Examinations, Evaluations of Sobriety, and Other Bodily Views and Intrusions Performed by Medical Personnel

CNICINST 3440.17, Navy Installation Emergency Management (EM) Program Manual

NAVMEDCOMINST 6710.9, Guidelines for Controlled Substances Inventory

NTRP 3-07.2.2, Force Protection Weapons Handling Standard Procedures and Guidelines

NTTP 3-07.2.1, Antiterrorism

OPNAVINST 1640.9 (series), Guide for the Operation and Administration of Detention Facilities

OPNAVINST 3100.6 (series), Special Incident Reporting (OPREP-3) Procedures

OPNAVINST F3300.53 (series), Navy Antiterrorism (AT) Program

OPNAVINST F3300.56, Navy Antiterrorism (AT) Strategic Plan

OPNAVINST 3440.17, Navy Installation Emergency Management Program

OPNAVINST 5530.13 (series), Department of the Navy Physical Security Instruction for Conventional Arms, Ammunition, and Explosives (AA&E)

OPNAVINST 5530.14 (series), Navy Physical Security and Law Enforcement Program

OPNAVINST 5585.2 (series), Department of the Navy Military Working Dog (MWD) Program

OPNAVINST 11200.5 (series), Motor Vehicle Traffic Supervision

SECNAV M-5210.1, Department of the Navy Record Management Program Records Management Manual

SECNAV M-5210.2 Department of the Navy Standard Subject Identification Code (SSIC) Manual

SECNAVINST 5212.5 (series), Navy and Marine Corps Records Disposition Manual

SECNAVINST 5300.29 (series), Alcohol Abuse, Drug Abuse and Operating Motor Vehicles

SECNAVINST 5430.107, Mission and Functions of the Naval Criminal Investigative Service

SECNAVINST 5500.29 (series), Use of Deadly Force and the Carrying of Firearms by Personnel of the Department of the Navy in Conjunction with Law Enforcement, Security Duties, and Personal Protection

SECNAVINST 5510.36 (series), Department of the Navy (DON) Information Security Program (ISP) Instruction

SECNAVINST 5530.4 (series), Navy Security Force Employment and Operations

SECNAVINST 5820.7 (series), Cooperation with Civilian Law Enforcement Officials

SECNAVINST S8126.1, Navy Nuclear Weapons Security Policy

WEBSITES

Naval Safety Center at: http://www.safetycenter.navy.mil/

INTENTIONALLY BLANK

GLOSSARY

antiterrorism (AT). Defensive measures used to reduce the vulnerability of individuals and property to terrorist acts, to include rapid containment by local military and civilian forces. (JP 1-02. Source: JP 3-07.2)

critical communications facility. A communications facility that is essential to the continuity of operations of the President or Secretary of Defense during national emergencies, and other modal points or elements designated as crucial to mission accomplishment. (NTRP 1-02.)

espionage. The act of obtaining, delivering, transmitting, communicating, or receiving information about the national defense with an intent, or reason to believe, that the information may be used to the injury of the United States or to the advantage of any foreign nation. Espionage is a violation of 18 United States Code 792–798 and Article 106, Uniform Code of Military Justice. (JP 1-02. Source: JP 2-01.2)

facility. A real property entity consisting of one or more of the following: a building, a structure, a utility system, pavement, and underlying land. (JP 1-02. Source: N/A)

force protection (FP). Preventive measures taken to mitigate hostile actions against Department of Defense personnel (to include family members), resources, facilities, and critical information. Force protection does not include actions to defeat the enemy or protect against accidents, weather, or disease. (JP 1-02. Source: JP 3-0)

force protection condition (FPCON). A Chairman of the Joint Chiefs of Staff-approved standard for identification of and recommended responses to terrorist threats against US personnel and facilities. (JP 1-02. Source: JP 3-07.2)

installation. A grouping of facilities located in the same vicinity, which support particular functions. Installations may be elements of a base. (NTRP 1-02.)

loss prevention. Part of an overall command security program dealing with resources, measures, and tactics devoted to care and protection of property on an installation. (NTRP 1-02.)

mobile vehicle inspection team. Security team consisting of two inspectors and a supervisor employed at entry control points as a part of a random antiterrorism measure program. (NTRP 1-02.)

Navy security force (NSF). Armed personnel regularly engaged in law enforcement and security duties involving the use of deadly force and unarmed management and support personnel who are organized, trained, and equipped to protect Navy personnel and resources under Navy authority. (NTRP 1-02.)

nonlethal weapon (NLW). A weapon that is explicitly designed and primarily employed so as to incapacitate personnel or materiel, while minimizing fatalities, permanent injury to personnel, and undesired damage to property and the environment. (JP 1-02. Source: JP 3-28)

physical security. That part of security concerned with physical measures designed to safeguard personnel; to prevent unauthorized access to equipment, installations, materiel, and documents; and to safeguard them against espionage, sabotage, damage, and theft. (JP 1-02. Source: JP 3-0)

Posse Comitatus Act. Prohibits search, seizure, or arrest powers to US military personnel. Amended in 1981 under Public Law 97-86 to permit increased Department of Defense support of drug interdiction and other law enforcement activities. (Title 18, "Use of Army and Air Force as Posse Comitatus"—United States Code, Section 1385.) (JP 1-02. Source: N/A)

random antiterrorism measures (RAMs). Random, multiple security measures that, when activated, serve to disguise the actual security procedures in effect. Random antiterrorism measures deny the terrorist surveillance team the opportunity to accurately predict security actions. Random antiterrorism measures strictly vary the time frame and/or location for a given measure. (NTRP 1-02)

restricted area. An area under military jurisdiction in which special security measures are employed to prevent unauthorized entry. (JP 1-02. Source: N/A)

risk. Probability and severity of loss linked to hazards. (JP 1-02. Source: JP 3-33)

risk assessment (RA). The identification and assessment of hazards (first two steps of risk management process). (JP 1-02. Source: JP 3-07.2)

risk management (RM). The process of identifying, assessing, and controlling risks arising from operational factors and making decisions that balance risk cost with mission benefits. (JP 1-02. Source: JP 2-0)

sabotage. An act or acts with intent to injure, interfere with, or obstruct the national defense of a country by willfully injuring or destroying, or attempting to injure or destroy, any national defense or war materiel, premises, or utilities, to include human and natural resources. (JP 1-02. Source: N/A)

security force. That portion of a security organization at a naval installation/activity comprising of active duty military, Department of Defense civilian police/guard, or contract guard personnel, or a combination, tasked to provide physical security and/or law enforcement. (NTRP 1-02)

security. Measures taken by a military unit, activity, or installation to protect itself against all acts designed to, or which may, impair its effectiveness. (JP 1-02. Source: JP 3-10)

terrorism. The unlawful use of violence or threat of violence to instill fear and coerce governments or societies. Terrorism is often motivated by religious, political, or other ideological beliefs and committed in the pursuit of goals that are usually political. (JP 1-02. Source: JP 3-07.2)

vehicle inspection team. Security force personnel who perform vehicle inspections as part of the installation access control program or are positioned throughout the base to conduct random vehicle inspections as part of a selective enforcement program. (NTRP 1-02)

LIST OF ACRONYMS AND ABBREVIATIONS

AA&E	arms, ammunition, and explosives
AECS	automated entry control system
AO	area of operations
AOR	area of responsibility
ART	armed response team
ASF	auxiliary security force
AT	antiterrorism
ATWG	antiterrorism working group
BOLO	be on the look out
BSO	budget submitting office
BUMED	Bureau of Medicine and Surgery (USN)
C2	command and control
CBRNE	chemical, biological, radiological, nuclear, and high-yield explosives
CBS	close boundary sentry
CCI	command criminal investigator
CCID	command criminal investigative division
CCTV	closed-circuit television
CDO	command duty officer
CENSECFOR	Center for Security Forces
CFR	Code of Federal Regulations
CID	criminal investigative division
CJCS	Chairman of the Joint Chiefs of Staff
CJIS	criminal justice information services (FBI)
CLEOC	consolidated law enforcement operations center

CNO	Chief of Naval Operations
CNOIVA	Chief of Naval Operations integrated vulnerability assessment
COG	commander of the guard
CPR	cardiopulmonary resuscitation
CQB	close-quarters battle
CW	cooperating witness
DBT	Design Basis Threat
DDD	drug detection dog
DOD	Department of Defense
DODD	Department of Defense directive
DODI	Department of Defense instruction
DON	Department of the Navy
DOT	Department of Transportation
DRRS-N	Defense Readiness Reporting System–Navy
DSCA	defense support to civil authorities
DSN	Defense Switched Network
DTRA	Defense Threat Reduction Agency
ECD	evidence/property custody document
ECP	entry control point
EDD	explosive detector dog
EM	emergency management
EMS	emergency medical services
EOC	emergency operations center
EOD	explosive ordnance disposal
ERT	evidence recovery team
ESS	electronic security system
EVOC	emergency vehicle operator course
FAP	Family Advocacy Program

FBI	Federal Bureau of Investigation
FPCON	force protection condition
FTO	field training officer (law enforcement)
GCC	geographic combatant commander
GSA	General Services Administration
HAZMAT	hazardous materials
HN	host nation
HSB	harbor security boat
HSPD	homeland security Presidential directive
IAW	in accordance with
ICO	installation commanding officer
ICP	incident command post
ICS	incident command system
ID	identification
IDS	intrusion detection system
IED	improvised explosive device
INIWIC	Inter-Service Nonlethal Individual Weapons Instructor Course
IR	incident report
ISIC	immediate superior in command
ISP	in-port security plan
ISSA	inter-Service support agreement
JP	joint publication
JSIVA	Joint Staff Integrated Vulnerability Assessment
LE	law enforcement
MA	master-at-arms
MANPADS	man-portable air defense system
MCM	Manual for Courts-Martial
MDMP	military decision-making process

MOA	memorandum of agreement
MOU	memorandum of understanding
MPD	military purpose doctrine
MSC	Military Sealift Command
MWD	military working dog
NAVFAC	Naval Facilities Engineering Command
NCIC	National Crime Information Center
NCIS	Naval Criminal Investigative Service
NDP	naval doctrine publication
NHTSA	National Highway Traffic Safety Administration
NIBRS	National Incident-Based Reporting System
NIJ	National Institute of Justice
NIMS	National Incident Management System
NKO	Navy Knowledge Online
NL	nonlethal
NLETS	National Law Enforcement Telecommunications System
NLW	nonlethal weapon
NRF	National Response Framework
NSF	Navy security force
NTRP	Navy tactical reference publication
NTTP	Navy tactics, techniques, and procedures
OC	oleoresin capsicum
OCONUS	outside the continental United States
OP	observation point
PA	public address
PAO	public affairs officer
PASGT	Personal Armor System for Ground Troop
PEO	Program Executive Office

PIO	police intelligence operations
POTUS	President of the United States
POV	privately owned vehicle
PPE	personal protective equipment
PS	physical security
RAM	random antiterrorism measure
RCA	riot control agent
RDT&E	research, development, test, and evaluation
REGCOM	regional commander
RFI	ready for issue
RIC	regional investigations coordinator
ROE	rules of engagement
ROI	report of investigation
RRP	rules, regulations, and procedures
RSO	regional security officer
RUF	rules for the use of force
SAM	surface-to-air missile
SD	surveillance detection
SDIC	surveillance detection information card
SecDef	Secretary of Defense
SHU	Scoville heat unit
SJA	staff judge advocate
SO	security officer
SOFA	status-of-forces agreement
SOP	standard operating procedure
SROE	standing rules of engagement
SSIC	standard subject identification code
SSS	strategic sealift ship

STAAT	security training assistance and assessment team
S&W	Smith and Wesson
TTP	tactics, techniques, and procedures
UCMJ	Uniform Code of Military Justice
UFC	Unified Facilities Criteria
UIC	unit identification code
USC	United States Code
USCG	United States Coast Guard
USN	United States Navy
USSS	United States Secret Service
VBIED	vehicle-borne improvised explosive device
VWAP	victim/witness assistance program
WC	watch commander
WMD	weapons of mass destruction

LIST OF EFFECTIVE PAGES

Effective Pages	Page Numbers
AUG 2011	1 thru 28
AUG 2011	1-1, 1-2
AUG 2011	2-1 thru 2-16
AUG 2011	3-1 thru 3-4
AUG 2011	4-1, 4-2
AUG 2011	5-1 thru 5-16
AUG 2011	A-1 thru A-16
AUG 2011	B-1 thru B-6
AUG 2011	C-1 thru C-14
AUG 2011	D-1 thru D-30
AUG 2011	E-1 thru E-12
AUG 2011	F-1 thru F-4
AUG 2011	G-1 thru G-4
AUG 2011	H-1 thru H-14
AUG 2011	I-1 thru I-8
AUG 2011	J-1 thru J-14
AUG 2011	K-1 thru K-6
AUG 2011	L-1 thru L-4
AUG 2011	M-1 thru M-8
AUG 2011	N-1 thru N-8
AUG 2011	O-1 thru O-22
AUG 2011	P-1 thru P-6
AUG 2011	Q-1 thru Q-4
AUG 2011	R-1 thru R-4
AUG 2011	S-1 thru S-8
AUG 2011	T-1 thru T-6
AUG 2011	U-1 thru U-4
AUG 2011	V-1, V-2
AUG 2011	W-1 thru W-6
AUG 2011	X-1, X-2
AUG 2011	Y-1 thru Y-4
AUG 2011	Z-1 thru Z-24
AUG 2011	Reference-1 thru Reference-4
AUG 2011	Glossary-1, Glossary-2
AUG 2011	LOAA-1 thru LOAA-6
AUG 2011	LEP-1, LEP-2

INTENTIONALLY BLANK

NTTP 3-07.2.3
AUG 2011

82825128R00176

Made in the USA
Lexington, KY
06 March 2018